# Noise Control in Buildings

# Noise Control in Buildings

Randall McMullan

OXFORD

BSP PROFESSIONAL BOOKS

LONDON  EDINBURGH  BOSTON

MELBOURNE  PARIS  BERLIN  VIENNA

DISTRIBUTORS

BSP Professional Books
A division of Blackwell Scientific
  Publications Ltd
Editorial offices:
Osney Mead, Oxford OX2 0EL
25 John Street, London WC1N 2BL
23 Ainslie Place, Edinburgh EH3 6AJ
3 Cambridge Center, Cambridge,
  MA 02142, USA
54 University Street, Carlton,
  Victoria 3053, Australia

First published 1991

Set by Setrite Typesetting Limited
Printed and bound in Great Britain
by The Alden Press, Oxford

Marston Book Services Ltd
PO Box 87
Oxford OX2 0DT
(*Orders*: Tel: 0865 791155
      Fax: 0865 791927
      Telex: 837515)

USA
Blackwell Scientific Publications,
Inc.
3 Cambridge Center
Cambridge, MA 02142
(*Orders*: Tel: (800) 759−6102)

Canada
Oxford University Press
70 Wynford Drive
Don Mills
Ontario M3C 1J9
(*Orders*: Tel: (416) 441−2941)

Australia
Blackwell Scientific Publications
(Australia) Pty Ltd
54 University Street
Carlton, Victoria 3053
(*Orders*: Tel: (03) 347−0300)

British Library
Cataloguing in Publication Data
McMullan, R. (Randall)
  Noise control in buildings.
  1. Noise. Control measures
  I. Title
  620.23

ISBN 0−632−02717−7

# Contents

# Introduction

Noise is one of the most common complaints about the built environment in which we live and work. The aim of this book is to help you build or to improve buildings with construction methods which make those buildings as quiet as possible.

Sound is a common experience in everyday life so it is easy for us to feel that we understand the effects of sound, just by instinct. In the same way that we know how to shut the door when it is cold or to open the plughole when we want to empty the bath.

But our instincts about sound are not always to be trusted. Test yourself on the following ideas.

- Noise sinks downwards through floors.
- Noise rises upwards through ceilings.
- Fibre glass quilt is a good sound insulator.
- Sheet glass is not a good sound insulator.

It is easy to agree with these statements but all of them are, in the main, *wrong*. The book explains why.

Incorrect ideas about sound wouldn't matter except that they are often used as the basis for action against noise. Effort and money are spent on special building techniques which are incorrect, or not as effective as they might be.

This book will help you put your efforts or your money into constructions which are more effective against noise. It will help you to understand the literature and the legislation about noise. I hope it also leads you to a quieter life.

My thanks to everyone who has helped to prepare this book and to the manufacturers who have supplied product information.

Randall McMullan

# Acknowledgements

The author and publishers thank the following organisations for permission to reproduce their material, shown accredited in the final section of the book.
Gyproc Insulation Limited, Runcorn, Cheshire, England.
Rockwool Limited, Bridgend, Mid Glamorgan, Wales.
Sound Attenuators Limited, Colchester, Essex, England.

# Using This Book

Part One of the book contains simple descriptions of sound, its behaviour and essential jargon. These sections will help you to understand the practical solutions to common noise problems which form the core of the book. Understanding the basic rules of sound may stop you taking a 'short cut' which wrecks the sound insulation.

In Part Two of the book there are practical construction details which can be used as models for building. Do read the nearby text which lists the essential features of each construction and give the reasons why it works. These practical details are supported by a product file showing a selection of commercial products used in noise control.

Part Three of the book gives more details of sound and noise theory, terminology and measurements, calculations and regulations. These technicalities are only needed for the more formal study of sound but they will help you to understand the literature and legislation about noise.

The Product File at the end of the book has manufacturer's installation drawings for a range of commercial products. These drawings provide further practical information and give a flavour of what is available.

# Part One
# Sound Ideas

This section of the book gives you the essential facts about sound and the effects it produces. The various topics of sound have certain specialised words associated with them and need some special systems of measurements. This information is intended to help you understand the practical advice given in Part Two of the book but you will also gain a good basic knowledge of sound, hearing, noise, insulation and acoustics.

# Nature of Sound

When a great tree in a remote forest falls over, and there is no one to hear it, does it make a sound? This question, which was pondered by philosophers of the past, now seems quite simple to answer; especially in this age of tape recorders. But the question raises thoughts that are still relevant: the main reason for our interest in sound is because it has a big effect on human life.

The study and the treatment of noise needs to involve the human reaction to sound but, initially, it is helpful to imagine a situation where we don't have ears or a sense of hearing. Instead, let's use neutral instruments which detect and measure sound. We could then ask about the nature of sound, and give an unbiased answer.

- Sound is a series of rapid variations in air pressure.

### Pressure

The changes in air pressure which cause sound are small, but the changes can be measured and they can be caused by all sorts of vibrating objects which push against the surrounding air. These vibrating objects might include strings, such as on a guitar or in your throat; panels, such in a loudspeaker or part of your wall; and moving masses of air, such as occur in pipes and jet engines.

### Speed

- The speed of sound in air is 340 metres per second.

The effect of sound moves forward through the air in the form of a vibrating wave motion which, just as the effect of waves in the sea, are seen to move. The speed of the waves, and therefore the speed of sound, is approximately 340 metres per second or 760 miles per hour.

The exact speed of sound varies with the temperature of the air but it is always slow enough to produce some noticeable effects in a large hall or arena. Surprisingly sound travels many times faster through denser materials such as water, wood and steel.

## Energy

- Sound is a form of energy.

The amount of energy involved in even the loudest sound is very small otherwise our ears would overheat! The energy content of a sound wave is often described by its *power* (energy given per second) and measured in watts (abbreviation: W).

We are interested in this energy content because of its importance in noise control. The basic method of stopping unwanted sounds is to convert the vibrational energy of the sound waves into other forms of energy such as frictional heat.

## Frequency

- Frequency is the rate at which the sound waves vibrate.

We interpret slow vibrations as *bass* notes and fast vibrations as *treble* notes.

Frequency is measured as vibrations per second called Hertz (abbreviation: Hz). Below about 20 Hz the waves are felt as a mechanical vibrations and above about 20 000 Hz, where we humans hear nothing, the vibrations are termed *ultra-sound*.

We are interested in sound frequency because the ear treats some frequencies differently to other frequencies and because most noise contains a mixture of frequencies.

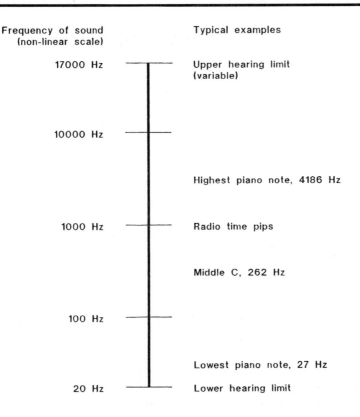

*Frequency of sound.*

# Hearing

There are minimum pressures which the average human hear can detect and maximum pressures which we can tolerate before damage occurs.

- *Loudness* of a sound depends on the pressure or the energy of the sound wave.

A sound of one fixed frequency can range between loud, with large vibrations and soft, with low vibrations.

This idea of loudness is complicated by the fact that the hearing system is most sensitive to sounds in the range 1000 Hz to 5000 Hz (or 1−5 kHz) which is in the range of shrieks and whistles. Sounds below 100 Hz, which we hear as 'bass' notes, need a lot more energy before we notice them as much as sounds in the sensitive frequencies.

This relative insensitivity of human hearing at low frequencies can be useful. Modern jet engines, for example, actually give out more sound energy than earlier types but they annoy us less because most of the sound energy is given out at low frequencies.

## Changes in loudness

A further complication of the hearing system is that when the sound energy is doubled it does not sound twice as 'loud'; in fact it is just noticeable. The energy needs to be increased about ten times before it seems twice as loud.

This may sound like good news for peace and quiet but, working in reverse, we need to reduce the *energy* of a troublesome sound to one tenth of its level for it to sound half as 'loud'; and to reduce the energy to one hundredth of the original level for it to sound about one quarter as loud. Noise control is going to involve large reductions in sound quantity or energy.

## Hearing damage

Hearing damage leading to deafness also depends on energy rather than on loudness, and this is unfortunate. For example, a tenfold increase in energy can increase the risk of hearing damage by ten but will only sound twice as loud. Similarly, a relatively small increase in loudness, such as increasing the volume of your personal stereo, can significantly damage your hearing.

In short, the sensitivity of our hearing system does not give us good protection against damage to the same system. The various forms of hearing loss are described in the technical section.

# Measurement

## Decibels

The uneven sensitivities of the human hearing system has led to the use of a decibel scale which is explained mathematically in the technical section. The simple energy or pressure measurements of sound are converted to sound level values in decibels (dB) which are easier numbers for humans to understand and relate to. Extra-terrestrial beings, or even your cat, might well prefer the unconverted values!

Sound levels in decibels start with a zero at the threshold of hearing, which is the weakest sound that the average human ear can detect. Typical effects of sound levels and changes in sound levels are shown in the illustrations. Remember that there is a distinct difference between a change in energy and a change in our idea of loudness.

A change in sound level of + or − 10 dB is a useful figure to remember as it makes difference of approximately twice as loud, or half as loud. We have to say 'approximately' as the experience also depends on individual hearing, on the background noise and on the exact frequencies involved. An increase in sound level of 20 dB (10 dB then another 10 dB) will seem four times louder.

For example, there may be a proposal to increase the average sound level of your environment from 60 dB to 70 dB. This seems a relatively small change, after all the scale runs from 0 to 140, but it will make the environment twice as noisy.

The same idea applies to reducing noise. If the manufacturers of a certain machine can reduce the sound level from 90 dB to 80 dB then the machine will sound approximately half as loud as before.

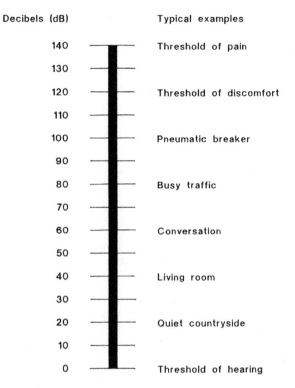

Decibels (dB)    Typical examples

140    Threshold of pain
130
120    Threshold of discomfort
110
100    Pneumatic breaker
90
80    Busy traffic
70
60    Conversation
50
40    Living room
30
20    Quiet countryside
10
0    Threshold of hearing

*Decibel scale of sound level.*

**Table of changes in sound level**

| Sound level change | Effect on hearing |
| --- | --- |
| ± 1 dB | negligible |
| ± 3 dB | just noticeable |
| + 10 dB | twice as loud |
| − 10 dB | half as loud |
| + 20 dB | four times as loud |
| − 20 dB | one quarter as loud |

## Sound meters

It is relatively simple to make instruments which can measure sound waves and display the results. Most *sound meters* detect sound with a microphone which converts the sound energy into electrical signals that can then be processed by electronic circuits.

Human hearing changes sensitivity with frequency, so if we measure noise at different frequencies, we will get different values. The most complete measurement of a particular noise, such as from your washing machine, is given by separåte readings of the sound level (in decibels) at standard bands of frequency and showing the results in a list or on a chart.

## A-scale

We often wish to measure a noise with a single measurement and this is done by allowing the sound level meter listen to all the frequencies at one time, just like the ear. An electronic circuit in the meter then emphasises the middle frequencies, just as the hearing system does. The results are usually quoted as dB(A) and the 'A' is an important tag telling us how the measurement was made.

Measurements in db(A) are widespread and convenient but you must remember that they are a compromise and tell us nothing about the frequency content of a noise. The ear can tell difference between a washing machine and a telephone buzzer because their sounds have a very different structure. But such different sounds may give identical readings in dB(A).

# Noise

In everyday speech the words *sound* and *noise* are used with roughly the same meaning. The word noise can have scientific meanings but there is also a simple environmental definition:

- Noise is unwanted sound.

Such a description takes account of the effect of a sound rather than the nature of a sound. Even the finest music is usually 'unwanted' in the middle of the night and so moves into the legal area of 'nuisance'.

## Tolerance

Remaining technical rather than legal, it is a fact that our acceptance of noise depends on more than one effect. In addition to individual sensitivities we need to consider the following factors.

### *Surroundings*

Acceptable levels of surrounding noise depend on what we are doing: sleeping, talking or hammering, for example.

### *Frequency content*

Some frequencies are more annoying than others. For example, we are more sensitive to high whines than to low rumbling noises.

## *Time duration*

A prolonged noise is more annoying than a short burst of noise.

These annoyance factors are easy to list but difficult to measure or to give numbers. Nevertheless we need systems of numbers, or *indices*, to describe the effect of noise. The numerical results of such measurements have to be matched to the results of surveys in which people rate their reaction to various levels.

## Noise measurement

Various systems of noise measurement have evolved for different types of situation. Some of the indices in use are listed here and explained further in Part Three: Technical Reference.

- *Equivalent continuous sound level*, $L_{eq}$ for environmental and industrial noise.
- *Noise Rating*, *NR* for machinery and services noise.
- *Traffic Noise Level*, $L_{10}$ for road noise.
- *Noise and Number Index*, *NNI* for airport noise.

# Sound Effects

Once you have generated a sound the effect immediately spreads beyond the source. In the simplest case, the sound waves spread out into the air. In most practical cases the sound waves meet obstacles, such as parts of a building, where the sound energy might be reflected, absorbed or transmitted. This section describes these effects and their practical consequences.

## Distance

As sound waves spread out from their source they *attenuate* or die away. This is because the total energy at the front of the wave remains almost constant but this fixed energy has to spread over an ever-larger area. The intensity or pressure of the sound energy measured at any point therefore decreases.

If a sound source acts like a perfect *point source* suspended in free space, then sound spreads out in the shape of a sphere. The intensity of the sound energy then varies in accordance with an *inverse square law*: which means that if you increase the distance by two then the sound intensity decreases to one quarter (the inverse of two squared).

When these energy figures are converted to the sound level scale of decibels, the reduction is 6 dB. Most practical sources of sound, such as engines or cars, are situated on a flat reflecting surface. All of the sound energy then radiates into a hemisphere rather that a whole sphere and the sound level decreases by only 3 dB.

- Doubling the distance from a single source of sound on a hard surface reduces the sound level by 3 dB.

If the source of the noise is a queue of cars on a busy road, or a similar *line source*, then the sound level reduces by only 1.5 dB when the distance from the source is doubled. Remember that you need to reduce the sound level by 10 dB in order to sound about half as loud.

The rather small reductions of 3 dB or 1.5 dB that you get when distance is doubled show that the distance from a noise has a relatively small effect over small distances, such as within a building. Eventually the repeated effect of distance does add up into a worthwhile reduction in sound level, but it usually means moving home!

## Sound mechanism

When sound meets a building surface such as a wall, floor, or ceiling there are several possible effects. Part of the sound energy is *reflected* back into the room, part is *absorbed* by the material and part is *transmitted* through into the next room. The amount of sound which is reflected, absorbed or transmitted depends on the type of material and the frequency of the sound. Figures are given in the technical section.

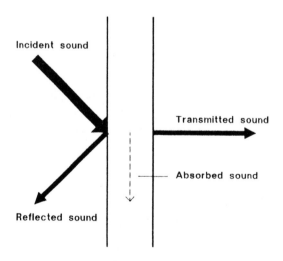

*Types of sound mechanism.*

## Sound absorption

Absorption occurs at a material when sound is not reflected from the surface and is not passed through the material. The sound energy has effectively 'got lost'.

Porous materials, such as glass wool, are good at absorbing high frequencies of sound. The pores need to be open and inter-connected so that sound waves are encouraged to enter further into the material. For heat insulation it doesn't matter whether the pores are open or closed. To absorb low frequencies of sound it is more effective to use suspended panels, such as plywood or panes of glass.

Much of the sound absorption in a room depends on porous materials, such as fabrics or glass fibre. The absorption provided by these materials has a big effect on the sound qualities (the acoustics) within the room but generally has little effect on the amount of sound passing in or out of the room (the sound insulation).

- Increasing the sound absorption in a room has little effect on the sound passing *between* rooms.

If, for example, you use *only* glass fibre to separate two rooms then that partition does not stop much sound passing between the rooms. On the other hand, the absorption of sound by the porous glass fibre would give different acoustic qualities to each room and cause them to sound rather 'dead'.

Although porous materials such as glass fibre are not good at directly stopping sound, you will see that these materials do feature in practical constructions for sound insulation. This is because absorption materials can break the sound linkage between parts of a construction such as two sides of a cavity. These absorbent materials also have useful properties of resilience and fire resistance.

The materials used for *heat* insulation are not therefore auto-matically suitable for *sound* insulation. When used correctly they help the sound insulation of some constructions but if they are just stuffed thoughtlessly into cavities the materials may do nothing much at all for sound insulation. See the practical con-structions shown in Part Two of the book.

## Sound reflection

Sound is reflected from a surface in the same way as light is reflected from a mirror, or as a ball is bounced off a wall. The angle made on the outward path matches the angle made on the inwards path. If the surfaces are curves then the reflection produces the effects shown in Part Three.

The reflection of sound has an important effect on the *acoustic* quality inside rooms by causing *reverberation* and *echoes*. Reflections along ducts and corridors also cause sound to travel from one part of a building to another.

Reflection can play a part in the control of noise. The direct sound from roadways and other noisy sites can be reflected away from sensitive areas, such housing sites. Earth banks or solid concrete walls make the best reflectors of outdoor sounds.

## Acoustics

When used technically, the word 'acoustics' is mainly concerned with *qualities* of sound such as evenness, clarity and fullness. Acoustics are less concerned with the sound insulation of a room, unless sound from outside the room is at an annoying level. The general requirements for good acoustics are summarised below.

- An even distribution of sound at adequate level.
- An appropriate reverberation time.
- An absence of unsuitable reflections such as echoes.
- An acceptable level of background noise.

The acoustics of a prominent auditorium, such as a concert hall, are given great thought at the design stage and plenty of money may be spent in getting the right results. Famous spaces, such as the circular Albert Hall in London, have also needed special techniques to 'tame' unwanted acoustic effects such as long echoes.

In most smaller halls and large rooms the final acoustic quality is usually left to chance. By applying a simple knowledge of acoustics we should be able to avoid building bad public environments such as echoing corridors or clattering restaurants. The same knowledge can also improve acoustic comfort in your home environment. For good acoustics in medium to large-sized rooms the general features listed below are of help.

- Irregular surfaces or decorations help to disperse the sound.
- Reflection by surfaces close to the source help to reinforce the sound.
- Absorption by surfaces far from the source help to prevent unwanted long reflections.

### Reverberation

Reverberation is the continuing presence of an audible sound after the source of sound has stopped. Most large rooms have a noticeable *reverberant field* which we can hear if we make a single sharp sound such as a hand clap.

Reverberation is caused by many rapid reflections between the surfaces of a room. These reflections are too rapid to be heard as separate sounds or echoes and, instead, the reverberations are heard as an extension of the original sound. Reverberation usually gives a better quality to sound, especially to music.

*Reverberation time* is the time taken for the reverberation to die away and this time is a useful factor for describing and controlling the acoustics of rooms. The reverberation time of

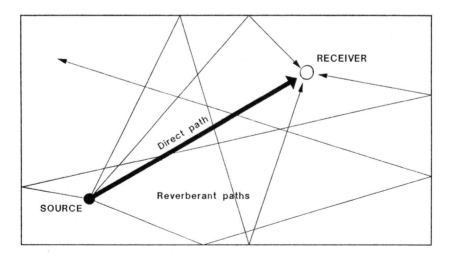

*Types of reverberation paths.*

existing spaces can be measured and the reverberation time of proposed spaces can be predicted with a knowledge of the room surfaces and methods such as Sabine's formula given in Part Three of the book.

The reverberation in a room depends on the absorption of the surfaces and in the distances between them. Typical reverberation times vary between a fraction of a second in small rooms to many seconds in large enclosures like cathedrals. The most suitable reverberation time for a room depends on the type of sound and the size of the room. Larger rooms need longer reverberation times.

In general, short reverberation times are needed for clarity of speech but cause music to sound 'dry' or 'unblended'. Longer reverbertion times enhance the quality of music but can cause speech sounds to overlap and become muddled.

## Sound transmission

The sound energy which passes through a partition, such as a wall, is said to be 'transmitted'. The various mechanisms and effects of sound transmission are important in explaining the sound insulation of various constructions shown in Part Two of the book.

The sound transmitted between different parts of a building is of two distinct types called *airborne* sound and *impact* sound. The type of sound depends on just what particular wall or floor is being considered.

- *Airborne* sources, such as voices, put the sound energy into the air from where it is passed into the structure of the building.
- *Impact* sources, such as footsteps or banging doors, occur *on* the structure of the building and put the sound energy straight into the structure.

The definition of whether a sound is airborne or impact depends on who is doing the listening. Footsteps, for example, will be heard as impact sound in the room below but the same footsteps will be heard as airborne sound in the room above. The two types of sound have to be measured separately.

One reason why it is important to know whether you are

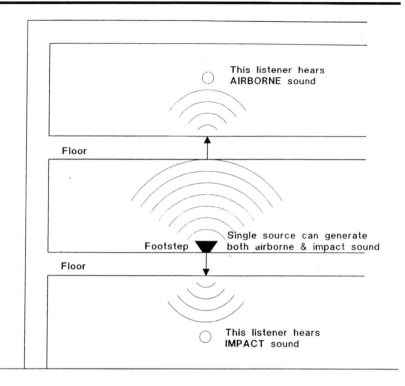

*Airborne and impact sound.*

dealing with airborne or impact sound is that methods of sound insulation are different for each type of sound. A heavy concrete floor, for example, gives good insulation against airborne sound but not against impact sound such as footsteps on the floor.

The Building Regulations regarding sound insulation also distinguish between airborne and impact sound. For example, the floors separating dwellings such as flats have to meet standards of insulation for both airborne and impact sound. Separating walls, on the other hand, only need to satisfy regulations against airborne sound as it is assumed that your neighbours won't walk on the walls!

## Flanking sounds

Even when you have successfully identified and reduced the sound passing directly through a wall or floor, you may be defeated by the sound which passes *around* that partition. Sound can flank around a partition by both airborne or structure-borne (impact) sound. Typical flanking routes are via common walls, suspended ceilings and open windows.

Flanking sound can ruin sound insulation and the practical constructions shown in Part Two of the book give details of the likely problems. If flanking sound cannot be stopped, such as when neighbours have windows open in the summer, then further sound insulation is of no use. It may be time to use the ultimate method of changing sound insulation − moving house!

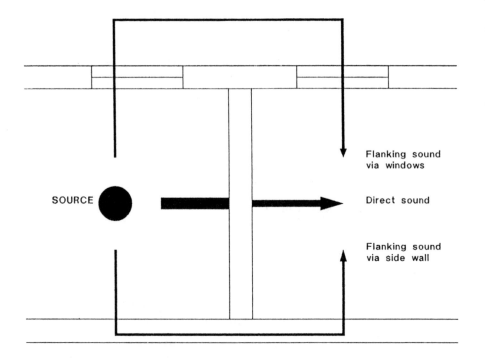

*Sound transmission paths.*

# Sound Insulation

Sound insulation is the main method of controlling the movement of sound within buildings. Remember that sound *insulation* is different to sound *absorption* which mainly controls the acoustic quality within rooms.

- Sound insulation is the reduction of sound passing *between* rooms or buildings.

Most of this sound is transmitted through partitions such as walls and floors as either *airborne sound* or *impact sound*, which were described in the last section. You need to identify the type of sound transfer in order to choose the best method of sound insulation.

The total effect of any sound insulation depends on how effectively you reduce all possible sound paths, direct and flanking; for both airborne and impact sounds.

### Sound insulation values

The effectiveness of airborne sound insulation can be measured by the difference in sound levels (in decibels) between each side of a partition. If, for example, a noise of 90 dB on one side of a wall is reduced to 40 dB on the other side then the *Sound Reduction Index* of the wall is $90 - 40 = 50$ dB.

Practical structures give different figures at different frequencies and full measurements must take this into account, as explained in the technical section. It is also common to compare sound reduction values at a single frequency, such as at 500 Hz.

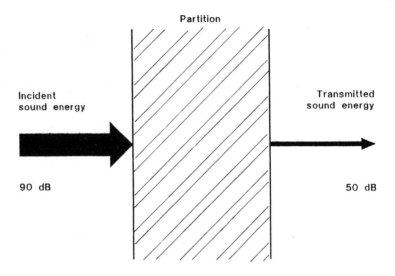

Sound reduction = 90 less 50 = 40 dB

*Sound reduction index.*


**Table of typical sound insulation values**

| Construction | Sound reduction index (average) |
| --- | --- |
| Brick cavity wall | 53 dB |
| Solid brick wall | 50 dB |
| Concrete blocks | 45 dB |
| Plasterboard on timber stud wall | 35 dB |
| Double window, 150 mm airspace | up to 40 dB |
| Single window, 6 mm glass, weather stripped | up to 25 dB |

Note: exact values of insulation depend upon the thickness and density of the materials and the construction details. See the details in the Practical Section.

Remember that a reduction in sound level of 10 dB sounds approximately half as loud as before. When listening through a wall, for example, different values of sound reduction have the effects shown in the Table of sound insulation effects.

**Table of sound insulation effects**

| Sound reduction (average) | Approximate effect on other side of partition |
| --- | --- |
| 20 dB | Normal speech is easily heard |
| 25 dB | Loud speech is easily heard |
| 35 dB | Loud speech is heard but not understood |
| 40 dB | Loud speech is just heard |
| 50 and above | Shouting is just heard |

## Principles of sound insulation

The following are the general principles which you should follow to achieve good sound insulation. The most appropriate principles depend on your particular problem and type of construction. The practical constructions shown in Part Two of the book often use a combination of construction methods and principles.

### Weight

Heavyweight parts of a building, such as brick walls, give good sound insulation. They can't pass on much sound into the next room because sound waves are vibrations and it is hard to vibrate a heavy wall.

The *Mass Law* says that the sound insulation of a single leaf partition rises in proportion to the mass of the construction. Theory and practice predicts that the sound insulation value increases by 5 or 6 dB for each doubling of the mass. For example, the average sound reduction of a brick wall increases from 45 to 50 dB when the thickness of bricks is doubled.

### Completeness

Airtight structures give good sound insulation. People don't readily accept the fact that small gaps in construction give large reductions in sound insulation. Seals around doors and windows

Poor ceiling fit

Poor panel fit

Poor door seal

Lightweight panel

Uncovered keyhole

Poor floor fit

Ceiling

Floor

*Typical insulation defects.*

need to be airtight, even the keyholes! Holes and ducts made for service pipes and cables are problems and some materials, such as blockwork, may be porous enough to pass sound.

In a complete partition, such as a wall, you only need a small area of lower insulation, such as a window, to greatly reduce the overall sound insulation of the complete partition. The components with the lowest insulation values, such as doors and windows, should be examined and improved before other measures are taken.

## Isolation

A minimum of physical contact between two surfaces or two rooms gives high sound insulation. An extreme example is to separate two rooms by an intermediate empty room. In practice we have to separate surfaces by much smaller air cavities or use resilient layers of material such as glass fibre or rubber. This isolation is easily ruined by transmission through rigid links, even by a single nail.

The air cavities within double-leafed constructions such as walls or double-glazed windows need to be as wide as possible to help prevent the stiffness of the air itself creating a link between the two surfaces. An absorbent layer within the cavity helps to prevent air linkages within the cavity and this is the main function of the mineral fibre often used in lightweight construction.

The effectiveness of most forms of sound insulation construction is seriously reduced at certain *critical frequencies* when the mechanical properties of the construction coincide with the wave properties of sound. These reductions, described further in the technical section, are called *resonance* or *coincidence* effects.

### Building Regulations

The Building Regulations, which legally require minimum standards of materials and construction in buildings, historically were concerned in making sure that buildings did not fall down on people and that they did not burn down. Many items in Building Regulations are a response to building disasters or bad practice.

The building codes of many countries pay attention to sound transmission between buildings and social surveys consistently show that the lives of many people are affected by the standard of sound insulation between themselves and their neighbours.

The Building Regulations for England and Wales have mandatory requirements to limit sound transmission through the walls and floors separating dwellings (houses and flats). These regulations, which can be taken as typical, are described further in Part Three.

# Strategies

This section summarises the general techniques for controlling noise in buildings. Part Two of the book shows these principles applied to practical constructions and Part Three gives details of theory and regulations.

## General principles

Previous sections have described how sound spreads and how it can be discouraged from spreading. Keeping this information in mind we can examine and control a noisy environment in three general categories:

(1) the source of the noise;
(2) the path that the sound takes;
(3) the receiver of the sound.

### *Sound source*

Factors to be considered include the following.

- Machines and processes. Improved technology and legal pressure can produce quieter sources of noise such as aircraft, heavy goods vehicles and construction equipment.
- Enclosures. Some noisy sources may be successfully contained near the source.
- Town planning. Busy roads and factories should be kept at a distance from housing.

- Siting. The spacing and staggering of houses can help avoid conflicts of noise between houses.
- Room layout. When houses or flats share walls the adjacent rooms should be compatible in their noise activities.

## Sound path

Factors to be considered include the following.

- Distance. As the distance from the source increases the received noise level slowly falls.
- Screens and barriers. Walls and embankments help deflect sound away.
- Surface treatment. Absorbent surfaces can help prevent sound travelling by reflections along corridors and ducts.
- Partition treatment. Insulated walls and floors help prevent sound transfer within a building.

## Sound receiver

Factors to be considered include the following.

- Building envelope. For external sounds the complete building acts as the receiver.
- Enclosure. An isolated area or room may be constructed within a building.
- Ear protection. The operators of some industrial processes may need to wear ear muffs.

You will notice that these areas of noise control have a logical order from start to finish. It is always best if you can tackle the problem nearer the start than the finish. For example, bedrooms in a block of flats should be designed adjacent to other bed-rooms; kitchens should be adjacent to kitchens. It is better to have silent neighbours than to have to wear ear muffs! Similarly, it is better to have the airport closed or made quieter than to insulate your home.

Being realistic, if we can't affect the sound source then we need to focus on the path of the sound or the receiver of the sound. The length of the sound path has to be greatly increased before it give a useful reduction in noise level. Nevertheless, building a long way from the airport is the ideal solution, even if it not useful to those already nearby. The effect of noise from busy roads is reduced by sinking the road in cuttings and by using earth embankments or stout walls. These reflective mechanisms work best for higher frequencies and will not eliminate the background rumble from a busy road.

Corridors and ventilating ducts often form sound paths which can be treated by absorbent surfaces. A single absorption is not particularly effective but the cumulative effect of absorbing multiple reflections is useful. Within a building the sound path is through the wall and floors. The rules of sound insulation, such as mass and air-tightness, should be applied to all partitions.

CORRIDOR or DUCT

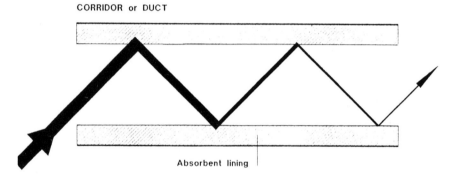

Absorbent lining

*Multiple absorption.*

The treatment of the sound receiver is last in the chain but it may be your only choice where the building itself is the receiver. Within a building you might have to isolate selected areas such offices in a factory, or a broadcasting studio. The final receiver is the human ear which may have to be protected from an inherently noisy process, such as drilling or cutting.

## Practical checklist

The following checklist summarises important practical measures which help give good sound insulation. These details are included in the sound-resistant constructions given in Part Two of the book.

## *Layout*

- Habitable rooms should face away from noisy roads.
- Adjacent buildings should be staggered to minimise adjacent wall areas.
- Adjacent rooms should have similar uses, such as sleeping.
- Entrance hallways and passages should be kept alongside party walls if possible.
- The shorter dimension rooms should be shared rather than the longer dimension ones.
- Adjacent windows should be small and separated by a wall extension if possible.

## *Services*

- Kitchen, bathrooms and services of flats should be vertically grouped above one another.
- Water supply and waste pipes should not run near living rooms and bedrooms.
- Boilers, pumps, phones and other vibrating devices should not be mounted on party walls.
- Ventilating plant and lift machinery should be isolated in areas with special insulation.
- Extractor fans should be mounted on rigid panels.

## *Doors and windows*

- Door closers should be fitted and correctly adjusted to prevent slamming.
- Door surrounds and keyholes should be sealed to a draught-proof standard.
- All glazing should use the thickest possible glass.
- Double glazing should use a large air gap.
- Openable windows should be sealed to a draughtproof standard.
- Ventilating ducts should be treated with baffles and absorbent surfaces.

## *Walls*

- Heavy bricks and blocks give good sound reduction.
- Bricks should be laid with frogs (indentations) upwards.
- All joints should be completely filled with mortar.
- Ultra-lightweight blocks designed for thermal insulation should not be used in party walls.
- The pores of concrete blocks must be sealed with plaster or cement-based paint.
- Cavity walls should use butterfly style of wall ties with the maximum spacing consistent with structural stability.
- Plasterboard partitions should have independent timber studs for different sides of the partition.
- Plasterboard should be doubled and overlapped.
- The airspace in a partition should be as wide as possible.
- Absorbent quilts can be used to increase isolation between the sides of a partition.
- The sound resistant construction of a party wall must continue into the loft space.
- All gaps at junctions with ceilings and floor must be sealed.
- Any gaps in a wall around joists or joist hangers should be sealed with mortar.
- Power outlets should not be installed back-to-back.
- TV and hifi speakers should not be attached to a party wall by brackets or shelves.

## *Floors*

- Heavy materials give good sound insulation.
- A resilient layer is needed to resist impact sound.
- Floating floors must not be connected to the structure beneath or to the surrounding walls.
- Skirting boards must not make a rigid connection with the floor.

# Part Two
# **Sound Solutions**

This part of the book gives practical details of common forms of construction and explains how their sound insulation works. It is a good idea to study first the features of those structures which give the best sound insulation and choose situations which can be compared to your own.

You will then have a better idea of what you need to do to provide effective sound insulation. Equally important: you will also know what *not* to do!

# Walls

Those walls in a house or flat which need to be sound-resisting, according to Building Regulations, follow common sense. They include the walls between a dwelling, such as a house or flat, and any other building; the walls between a habitable room and any other part of the same building which is not part of the same dwelling; and the walls between a refuse chute and any rooms in a building.

These requirements are only for insulation against airborne sound like voices, not against impact sounds on the wall like hammering, noisy pipes or vigorous bedsteads. For better or worse, Building Regulations do not yet require party walls to be insulated against impact sound.

If annoyance is being caused by impact sound, such as noisy pipes, it can be reduced significantly by using isolation techniques, similar to those shown in the floor constructions. Resilient layers, like the glass fibre shown in the timber frame walls, will be effective but it is much simpler and cheaper to isolate the source of the vibration, such as the pipe mountings. Isolation effects will be ruined if there are solid sound bridges between the partitions: pipes, power sockets, wall ties, nails are all possible offenders.

In countries such as the British Isles, brick or stone is a traditional building material for dividing walls and these materials provide good sound insulation. This is a happy accident resulting from a shortage of timber and a regard for fire precautions, rather than a historical desire for high sound insulation.

# SOLID MASONRY WALLS

Masonry is a general term for brickwork, concrete or stonework which are often used for those party walls which separate different dwellings or offices. By happy accident, masonry walls generally provide good sound insulation and good fire resistance.

The traditional party wall construction of one brick thickness (measuring by the long side of a brick) with plaster on both sides gives a standard of airborne sound insulation which meets the requirements of Building Regulations. Surveys of people living in houses separated by this type of wall indicate that, at this standard of insulation, most of them are not 'unduly' annoyed by their neighbours.

Typical performance figures for a party wall give an average sound reduction value of around 50 dB, which is like reducing the noise level of noisy truck in the street to that of a quiet office. At this level of insulation you will only just hear shouting from the other side of the wall.

In most practical cases this sound insulation value of 50 dB is reduced by flanking sounds which find paths around junctions and through windows. All the same, this is as good a result as you can build without special construction techniques.

### How it works

Refer to the construction details shown in the diagram for *solid masonry walls*.

- The *weight* of the masonry provides resistance to airborne sound.
- The *weight* of the plaster provides resistance to airborne sound.
- The *seal* of the plaster provides resistance to airborne sound.
- The wall must be *airtight* in its construction and at its junctions.

### How it is built

The weight of the wall is usually specified as a weight per square metre of wall area, also called surface density, in units of $kg/m^2$.

215 mm

**BRICKWORK WITH PLASTER**

Plaster, 13 mm

Brickwork, 215 mm

Plaster, 13 mm

CARE
Lay bricks with frogs upwards.
Fill joints completely with mortar

Cross section
(not to scale)

**BLOCKWORK WITH PLASTER**

Plaster, 13 mm

Blockwork, varying thickness
to achieve 415 kg per sq metre

Plaster, 13 mm

CARE
Lay blocks with full thickness
across wall.
Use correct weight of blocks

Cross section
(not to scale)

*Solid masonry walls.*

The necessary thickness of the wall therefore depends on the
type of material the wall is built from. As an obvious example,
you need a greater thickness of lightweight concrete blocks than
heavy engineering bricks in order to achieve the same value in
$kg/m^2$.

## *Brickwork*

For solid brickwork, plastered both sides, the minimum weight is specified as 375 kg/m$^2$. A typical density for a common brick with mortar is 1800 kg/m$^3$. Therefore the minimum thickness you need of such brickwork is the fraction 375/1800 of one metre, which equals 0.208 m or 208 mm. A brick with a standard length of 215 mm therefore meets the requirement and the plaster will add further useful weight.

Another way of meeting the requirement is to check the average weight of one dry brick. In order to meet the required surface weight of 375 kg/m$^2$ the brick must have a minimum weight of 1.81 kg. This assumes that the brick has standard dimensions of 65 by 103 by 215 mm, is laid with frog (indentation) upwards and is plastered both sides. Brick manufacturers supply suitable density and weight information in their data sheets.

## *Blockwork*

For concrete blockwork, plastered both sides, the minimum surface weight is specified as 415 kg/m$^2$. If the blocks measure 100 by 215 by 440 mm then each block will need to weigh 16.7 kg. If the blocks measure 215 by 215 by 440 mm then each block will weigh 36 kg. The manufacturers' data sheets for concrete blocks give suitable information.

These are relatively heavy concrete blocks compared with the ones in common use for thermal insulation but they look the same when heaped up on site. You should identify the heavy blocks, with paint perhaps, and make sure that they are used in the right place. Busy bricklayers don't always appreciate the importance of weight for sound insulation.

The plaster should be at least 12.5 mm thick on each face of the wall and is important for making the wall airtight, especially when concrete blocks are used.

## *Joints*

Bricks must be laid in a brickwork bond (pattern) which includes headers (ends) showing. In other words, a significant number of bricks need to run right through the wall without gaps. Concrete blocks must also be laid in a bond in which the blocks extend for the full thickness of the wall without gaps in the middle.

The joints between bricks or blocks must be *completely* filled with mortar. If the bricks have frogs (indentations) then the frogs must be laid pointing upwards and they must also be completely filled with mortar.

You can achieve adequate structural bonding between bricks and blocks with a minimum of mortar so bricklayers don't usually see the need to use more mortar. They may also think, quite correctly, that it is going to be difficult for anyone to check on what they have done!

## *Junctions*

The sound-resisting wall must be adequately tied to the external walls. With masonry external walls there is usually bonding between the brick or blockwork of each wall. Otherwise the external wall, brick or timber-frame, must be firmly butted against the sound-resisting wall and secured with ties at a minimum vertical spacing of every 300 mm.

The openings in an external wall on either side of a party wall should be separated by at least 650 mm.

## *Roof spaces*

Party walls usually continue into a roof space where they need to resist flanking sound. The weight of the wall may be reduced but it must remain airtight. Junctions need to be sealed and lightweight (therefore porous) concrete blocks need to be sealed with plaster or cement paint.

## Variations

*Cavity walls* built of bricks or blocks give, surprisingly, a negligible increase in sound insulation. This is because the rigid wall ties which cross the cavity, and are essential for structural purposes, are also efficient transmitters of sound. The practical sound insulation of a masonry cavity wall therefore remains dependent on its total weight, which is usually the same as the solid wall built with the same number or bricks or blocks.

*Half-thickness* masonry walls may be used as internal partitions, and are often found in older houses. They are usually 75 to 100 mm (3 to 4 inches) in thickness and are solid when you tap or drill them. If built of brick these internal walls provide a sound reduction value not much less (5 dB) than a standard party wall. This value decreases slightly when concrete blocks are used.

Precast *concrete panels* behave like a masonry construction of similar surface weight, provided that the panels are butted together and that all the gaps are sealed.

# MASONRY AND PANEL WALLS

The masonry walls described in the previous section meet the sound insulation requirements of the building regulation by virtue of their heaviness. If the masonry construction is not heavy enough, such as with lightweight concrete blocks, then you can increase the sound insulation by adding a separate panel to the wall, as shown in this section.

This type of construction can give an airborne sound insulation value of around 50 dB, the same as for a heavy brick party wall, provided that the flanking sound paths are limited. To meet full party wall standards this type of construction should only be used with a concrete ground floor.

## How it works

Refer to the construction details shown in the diagram for *masonry and panel walls*

- The *weight* of the wall provides resistance to airborne sound.
- The *weight* of the panel provides resistance to airborne sound.
- The physical *isolation* of the two parts of the wall provides resistance to airborne sound.
- The wall must be *airtight* in its construction and at its junctions.

## How it is built

The illustration for the *masonry and panel walls* shows a masonry core of brick or blockwork together with an independent panel built from double plasterboard or from manufactured cellular partition material. The panel can be installed on either side of the masonry core.

### Brickwork or blockwork

For bricks and concrete blocks, the surface weight of the masonry needs to be at least 300 kg/m². This weight is achieved by most

MASONRY CORE WITH LIGHTWEIGHT PANEL

Brickwork or Blockwork, 300 kg per sq metre

Air space, 25 mm minimum

Cellular panel, 18 kg per sq metre or
Plasterboard, 30 mm

Supporting framework (optional)

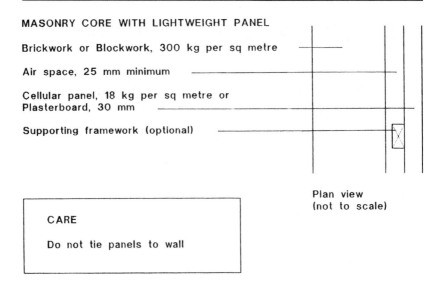

Plan view
(not to scale)

CARE

Do not tie panels to wall

JUNCTION WITH INTERMEDIATE FLOOR

Masonry core wall

Panel, fixed to batten on floor

Floor

Joist hanger
Timber joist
Timber blocking between joists

Ceiling

Tape seal at junction

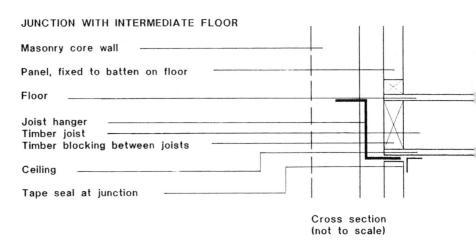

Cross section
(not to scale)

*Masonry and panel walls.*

bricks and by blocks (sized 100 by 215 by 440 mm) if they each weigh at least 14.2 kg when dry.

If lightweight aerated concrete blocks are used the surface weight must be at least 160 kg/m$^2$. This weight is achieved when each block weighs 5.8 kg for a block size of 100 by 215 by 440 mm; or when each block weighs 13.6 kg for a block size of 215 by 215 by 440 mm. The data sheets from brick and block manufacturers supply appropriate tables of weights.

## *Panels*

The lightweight panel is built on either side of the wall and fixed to timber battens on the floor and the ceiling but not fixed to the masonry core. Partition board, made of two sheets of plasterboard joined by a cellular core, should have a surface weight of at least 18 kg/m$^2$. The joints between panels are sealed with tape so as to avoid air paths.

An alternative panel construction is two sheets of plasterboard with a total thickness of 30 mm. The two layers of plasterboard, with staggered joints, can be bonded together or attached to a supporting framework. The joints between the plasterboard sheets are staggered and sealed.

## *Junctions*

The joists of intermediate floors are attached to the masonry core of the wall with joist hangers. When the joists run at right angles to the wall the spaces between each joist is blocked in with a timber batten.

Where loadbearing partitions meet the wall thery are carried through the panel and fixed to the masonry core, separated by a padding of mineral fibre quilt. Non-loadbearing partitions are tightly butted against the panel. All joints and gaps between walls or between walls and ceilings are sealed with suitable mastic, tape or coving.

The openings in an external wall on either side of a party wall should be separated by at least 650 mm.

## *Roof spaces*

Where party walls continue into a roof space the weight of the wall may be reduced but it must remain airtight. Junctions need to be sealed and lightweight (therefore porous) concrete blocks need to be sealed with plaster or cement paint.

## Variations

The specifications of this wall can be relaxed and still achieve respectable degrees of sound insulation which will be adequate for many situations. Within a dwelling, for example, the flanking sounds through doors and other areas make it pointless to build internal partitions to full party wall standard.

The sound insulation and construction techniques used for this wall can also be used for upgrading the sound insulation of existing masonry walls.

# TIMBER FRAME WALLS

It is possible to build a timber-framed wall which gives the same high standard of sound insulation and fire resistance as a masonry party wall. The design and construction of lightweight walls, however, need rather more care than the relatively effortless sound insulation gained from heavy brickwork.

The construction considered here is designed to meet party wall regulations and give a sound insulation of 50 dBs, which is the same as a brick or block party wall. In practical situations, this level of insulation is reduced by the flanking paths available for sound through windows or doors, especially within the same dwelling.

A simple timber-framed wall with plasterboard on both sides gives a sound reduction of about 35 dB before improvements to sound insulation. At this level of insulation, which is considered adequate for partitions inside a home, you shouldn't hear normal conversation through the wall but you will be able to hear loud speech.

For the moment we are only considering airborne sound, such as voices; the impact sounds caused by pipes, WCs, washing machines or other equipment need different treatment.

## How it works

Refer to the construction details shown in the diagram for *timber frame walls*.

- The *weight* of the plasterboard provides resistance to airborne sound.
- The physical *isolation* of the two frames provides resistance to airborne sound.
- The *width* of the air space provides resistance to airborne sound.
- The acoustical *isolation* of the mineral fibre provides resistance to airborne sound.
- The wall must be *airtight* in its construction and at its junctions.

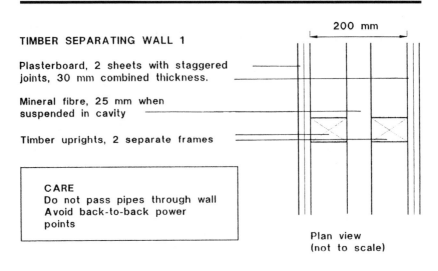

TIMBER SEPARATING WALL 1

Plasterboard, 2 sheets with staggered
joints, 30 mm combined thickness.

Mineral fibre, 25 mm when
suspended in cavity

Timber uprights, 2 separate frames

CARE
Do not pass pipes through wall
Avoid back-to-back power
points

Plan view
(not to scale)

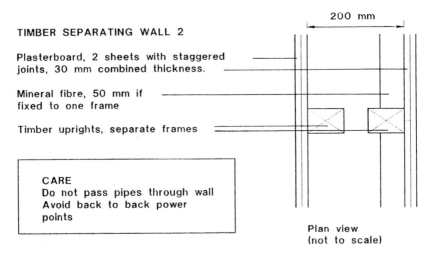

TIMBER SEPARATING WALL 2

Plasterboard, 2 sheets with staggered
joints, 30 mm combined thickness.

Mineral fibre, 50 mm if
fixed to one frame

Timber uprights, separate frames

CARE
Do not pass pipes through wall
Avoid back to back power
points

Plan view
(not to scale)

*Timber frame walls.*

## How it is built

The illustration for *timber frame walls* shows a separating wall
formed by two timber frames which are built independently of
one another. The timbers are typically 75 by 50 mm in cross

section and placed so that the outside edges of the two frames are separated by at least 200 mm.

The vertical timbers (studs) are spaced at suitable distances, such as 450 mm or 600 mm, which coincide with the edges of the plasterboard sheets. The studs are attached to separate floor and ceiling plates. Cross timbers (noggins) are also installed, with the help of skew nails, where the plasterboards need support.

### Absorbent curtain

Mineral fibre is the generalised name for blankets of resilient material such as glass fibre, or glass wool. For maximum absorption of sound the texture of the fibre must have an open weave rather than closed cells. There are other resilient materials which could perform similar functions but mineral-based products, like glass fibre, also offer fire resistance and stability.

The absorbent curtain between the frame uses mineral fibre with a minimum density of 12 $kg/m^3$. The curtain is at least 25 mm in width if it suspended in the cavity or 50 mm if it is attached to one of the frames.

### Plasterboard

The term 'lightweight' wall is only relative, as you will agree after moving plasterboard about. The shear weight of the plasterboard on each side of this wall is an important feature of the specification and the total thickness of plasterboard needs to be 30 mm. Don't use lightweight sheeting materials.

The plasterboard is nailed to the frames. The two layers of plasterboard are staggered so that they are fixed to alternate studs and the joints don't coincide.

### Junctions

All junctions between the sound-resisting wall and other walls, floors and ceilings need attention to detail so as to block air paths. Appropriate coving, skirting boards and sealants should be used. When the joists of an intermediate floor are at right angles to the wall, use timber blocking in the spaces between the joists.

If, for structural reasons, the two frames of the sound-resisting wall need to be connected then use metal straps fixed below ceiling level and with a minimum horizontal spacing of 1.2 m. If fire stops are needed in the wall then they should be flexible like mineral wool or, if rigid, they must fixed to one frame only.

## *Airgaps*

It is much easier to make holes in timber frame walls than in masonry walls so fitters of pipework and cables, including house-holders, are therefore more inclined to pass items right through the walls in more places. All such installations leave airgaps and many rigid fittings, such as pipes, make acoustic bridges which ruin the isolation of the frames.

If airgaps do exist then they should be sealed; the type of mineral wool used in the wall cavity and patching plaster is often satisfactory. When power points are set in the cladding on one side of the wall they need to be surrounded by a similar thickness of plaster to the rest of the wall. In effect you make a recess of doubled plasterboard. Don't have power points back-to-back in the same position on each side of the wall. It might be an easy option for the electrician but it helps ruin sound insulation.

## *Roof space*

Where party walls extend into a roof space then both frames and claddings of the wall may be carried through to the underside of the roof. Alternatively, the cavity can be closed at ceiling level, avoiding rigid connections between them. The party wall can then be continued to the roof with a minimum of 25 mm cladding on each side. All junctions need to be kept airtight.

## Variations

It is possible to relax the above specifications and still achieve respectable degrees of sound insulation which will be adequate for many situations. Within a dwelling, for example, the flanking sounds though doors and other areas make it pointless to build internal partitions to full party wall standard.

Some of the sound insulation and construction techniques used for this type of wall can also be used for upgrading existing walls, as shown in a later section.

*Staggered studwork* is a method of imitating the structure of two completely separate frameworks within a single framework. The sheeting on each side of the is attached to separate alternate studs.

The essential technique is to keep the two sides of the wall isolated by a resilient material and to prevent rigid contact between them. The relatively narrow airgap means that the two sides of the wall are not acoustically isolated as effectively as when using a wider gap.

*Mineral fibre* just stuffed inside a standard timber-framed/ plasterboard increases the sound insulation by about 5 dB. At some frequencies this improvement in sound insulation will be hard to detect and, unless you have want to get rid of mineral fibre, it is worth going to a little more trouble.

*Metal channels* may be used instead of timber studs. The sound insulation of the various office partition systems depends on the thickness of the plasterboard used. The sound insulation using metal studs is the same or slightly above similar con- struction using timber studs.

Some commercial systems of *dry lining* use plasterboard panels to which mineral wool is bonded. The mineral wool is used to acoustically separate the panel from a lightweight masonry wall. If such dry lining is fitted on both sides of the masonry wall it can give a sound insulation performance similar the masonry and panel wall shown in this section.

# REMEDIAL WALL

People often wish to improve the sound insulation of a wall between them and their neighbours. If you live in a flat formed by the subdivision of a large house then it is quite possible that the sound insulation of your separating wall is of a lower standard than that required by modern regulations, and less than what is acceptable.

The remedy described here assumes that you only have access to your side of the wall and uses a new and independent wall built in your room. Once again, a quieter room is a smaller room!

The construction of the independent wall will improve the sound insulation against airborne sound by between 5 and 10 dB, depending on the frequencies contained in the noise. An improvement of 5 dB is noticeable and an improvement of 10 dB roughly corresponds to a halving in loudness.

Before starting this type of remedial treatment you should try to ensure that most of the troublesome noise is actually coming through the wall and not via flanking routes at the ceiling and wall junctions, or through windows. If you and your neighbour both have windows which are close to one another then you may be disappointed by the upgraded wall unless your window is also upgraded.

## How it works

Refer to the construction details shown in the diagram for *remedial walls*.

- The *weight* of the plasterboard provides resistance to airborne sound.
- The physical *isolation* of the two parts of the wall provides resistance to airborne sound.
- The *width* of the air space provides resistance to airborne sound.
- The acoustical *isolation* of the mineral fibre provides resistance to airborne sound.
- The wall must be *airtight* in its construction and at its junctions.

INDEPENDENT WALL LINING

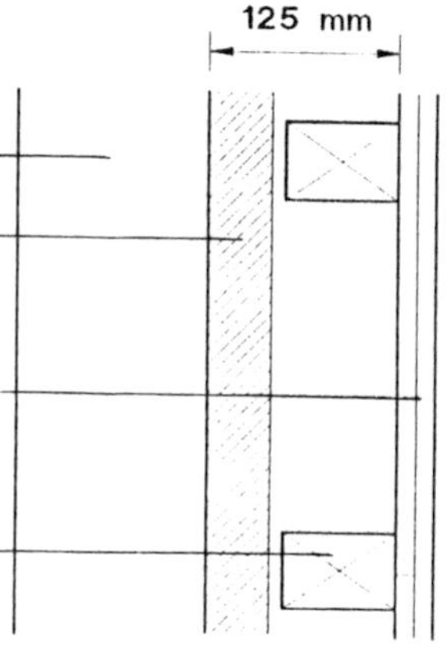

Existing wall

Mineral fibre, 25mm min
loosely hung

Plasterboard, 2 sheets with staggered
joints, 30mm combined thickness

Timber upright

125 mm

Plan view
(not to scale)

CARE

Do not tie new wall to the
existing wall

TIMBER FAMEWORK

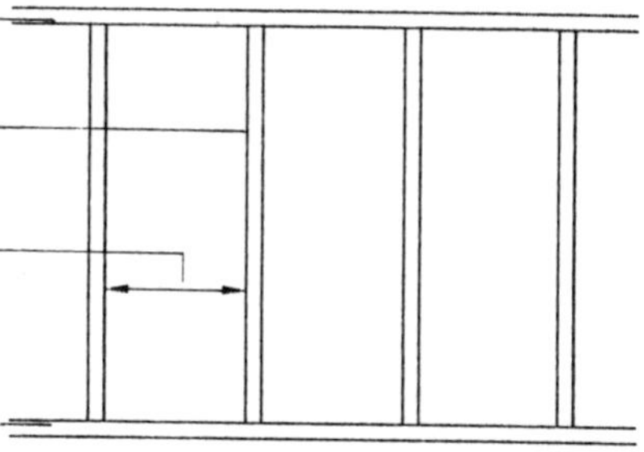

Ceiling plate

Timber uprights

Spacing to suit plasterboard

Floor plate

Front view of frame
(not to scale)

*Remedial walls.*

## How it is built

The *remedial wall* diagram shows a new timber frame wall which
is built independently of the original wall. The timbers are typically
75 by 50 mm in cross section and placed so that the face of the
plasterboard will be at least 125 mm from the existing wall. You

can attach the new wall to the side walls but never to the original wall that is being improved.

The vertical timbers (studs) are spaced at suitable distances, such as 450 mm or 600 mm, which coincide with the edges of the standard plasterboard sheets. The studs are attached to timber plates attached to the floor and ceiling plates. Cross timbers (noggins) are installed between the studs in places where the edges of plasterboards need support.

### Absorbent curtain

The curtain of mineral fibre, such as glass fibre or rock wool, is loosely hung in the cavity between the two parts of the wall. A 25 mm thickness can be placed against the existing wall face and held by a batten on the ceiling. Alternatively a thicker quilt, like that used for thermal insulation, can be hung inside the framework.

### Plasterboard

The shear weight of the plasterboard in the new wall provides a lot of sound insulation and the total thickness of plasterboard needs to be 30 mm. Don't use lightweight sheeting materials. The plasterboard is nailed to the frames. The two layers of plasterboard are staggered so that they are fixed to alternate studs and the joints don't coincide.

### Junctions

All junctions between the sound-resisting wall and other walls, floors and ceilings need to be carefully sealed in order to block air paths. Use appropriate coving, skirting boards and mastic sealants.

### Variations

Commercial systems of *dry lining* are available, using dense factory-made boards and a layer of mineral wool. If installed correctly they can give worthwhile improvements in sound insulation.

# Floors

Floors which separate dwellings, such as different flats in a block, require the highest standards of sound insulation that we meet in regular construction activities. Such floors have to achieve good insulation against airborne sound, like voices, and also against impact sound, such as footsteps.

Modern purpose-built flats have to satisfy Building Regulations and will have sound-resisting floors between flats. But the internal floors inside one dwelling, such as a house, have never had special treatment for sound insulation; presumably on the grounds that sources of sound within your own household are under your control!

If a house is converted into flats the sound insulation of the intermediate floors needs upgrading, although there are many conversions where this process has been neglected or has been poorly done. Residents in such flats are often interested in better sound insulation and details of remedial treatments are shown at the end of this section. It is also helpful to look at the other floors and to understand how the floor should have been built!

The sound insulation of all floors is affected by sounds travelling through the adjacent walls, or through openings such as cracks and windows. As the level of sound insulation is increased these flanking sounds become dominant and it becomes unproductive to improve the floor unless the other areas can also be improved.

# CONCRETE FLOOR WITH SOFT COVERING

A concrete floor has the benefit of mass, whether the concrete is poured on site or used in the form of concrete beams. This heaviness automatically gives a good standard of sound insulation against airborne sound, similar to that of a party wall. But the concrete provides only slight insulation against impact sound and a soft resilient layer needs to be incorporated in the floor.

The floor constructions shown here satisfy the requirements of the Building Regulations for floors between dwellings, both for airborne and impact sound. Surveys of people living with such floors have found that such standards of sound insulation are generally regarded as reasonable for most of the time.

### How it works

Refer to the construction details shown in the diagram for *concrete floors 1*.

- The *weight* of the concrete base provides resistance to airborne sound.
- The floor must be *airtight* in its construction and at its junctions.
- The physical *isolation* of the resilient covering gives resistance to impact sound.
- The isolation must be *complete* and not reduced by rigid links or flanking transmissions.

### How it is built

The illustration for *concrete floors 1* shows a concrete floor base whose weight is specified by a surface density of 365 kg/m$^2$. The actual thickness of the floor therefore depends on the final density of the concrete that is used. If the concrete has a density of 2400 kg/m$^3$, a typical figure for plain concrete, then the thickness of concrete you need is the fraction 365/2400 of 1 metre, which equals 0.152 metres.

CONCRETE FLOOR WITH SOFT COVERING

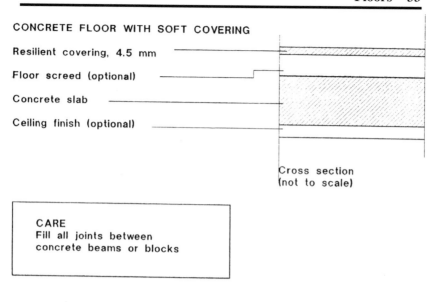

Resilient covering, 4.5 mm

Floor screed (optional)

Concrete slab

Ceiling finish (optional)

Cross section
(not to scale)

CARE
Fill all joints between
concrete beams or blocks

JUNCTION WITH WALL

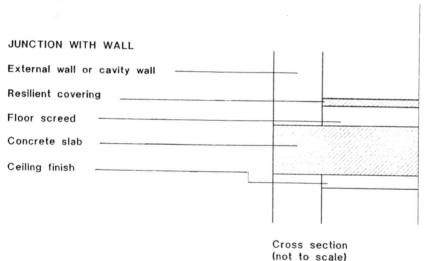

External wall or cavity wall

Resilient covering

Floor screed

Concrete slab

Ceiling finish

Cross section
(not to scale)

*Concrete floors 1.*

This thickness of around 150 mm (approximately 6 inches) is typical of floor systems. The figure of 365 kg/m$^3$ can include any floor screeds, any shuttering and any ceiling finish bonded to the floor so the thickness of the floor slab can usually be less then 150 mm.

The resilient layer, or soft covering, is often a carpet or vinyl floor covering. The total thickness of the layer must be 4.5 mm, which can include backing, and only the base has to be resilient. The test of a resilient material is that it returns to its original thickness after it has been compressed.

## *Junctions*

Where a solid floor meets an external wall or a cavity separating wall, the base of the floor, but not the screed, should be passed through the wall. This applies even when the floor beams are running parallel to the wall and the junction is not needed for structural support. If the floor is constructed from prefabricated concrete units such as hollow concrete beams, then the first joint should be at least 300 mm from the face of the wall.

If the floor meets an internal solid wall weighing less than 355 kg/m$^2$ the floor base should be passed through the wall. If the wall is sound resisting or heavier than 355 kg/m$^2$ there is a choice of passing the wall or the floor base through. When the wall is passed through it is important to tie the floor base to the wall and to seal the joint with grout.

For any type of external wall which has openings 20 per cent or less of the external wall area, then the weight of the external wall, or inner leaf of a cavity wall, should have a surface density of at least 120 kg/m$^2$. This figure can include plaster but not dry lining. Where the openings are more than 20 per cent of the external wall area, there are no requirements in the building regulations.

The overall performance of the floor is affected by possible flanking sound via construction details on either side of the floor. If a pipe has to penetrate the floor then it should be contained in ductwork made from a board material with a weight of at least 15 kg/m$^2$. All pipework inside the duct should be wrapped in a minimum 25 mm of mineral fibre.

## Variations

*Soft coverings* which are thicker and more resilient than the minimum specified in the regulations will improve the insulation against impact sound such as footsteps. Thick carpet and a springy form of underlay are helpful.

*Floating layers* are a special construction feature which increases the sound insulation and usually allows a reduction in the weight of the floor. For details, see the constructions detailed in the following sections of the book.

# CONCRETE FLOORS WITH FLOATING LAYER

A *floating floor* is, as its name suggests, constructed so that there are two separate layers in the floor. These layers are physically isolated from one another by the use of resilient materials, such as mineral fibre and special forms of expanded plastic.

The top half of the floor just rests on the base of the floor and is held there by the force of gravity. There must be no rigid links and even a single nail will act as a sound bridge between the two parts of the floor, just as a small piece of wood passes all the sound from a violin string to the violin case.

When built correctly, the floating floor constructions shown here satisfy the requirements of the Building Regulations for airborne and impact sound. Surveys have found that such standards of sound insulation are regarded as reasonable by most households for most of the time.

The use of a floating layer gives good standards of sound insulation while using a lighter weight of concrete than in the floors described previously. This reduction in required weight allows you to reduce the thickness of the concrete or to use lighter concrete beams.

### How it works

Refer to the construction details shown in the diagrams for *concrete floors 2* and *concrete floors 3*.

- The *weight* of the concrete base provides resistance to airborne sound.
- The *weight* of the floating layer provides resistance to airborne sound.
- The acoustical *isolation* of the floating layer also provides resistance to airborne sound.
- The floor must be *airtight* in its construction and at its junctions.
- The physical *isolation* of the floating layer gives resistance to impact sound.
- The isolation must be *complete* and not reduced by rigid links or flanking transmissions.

CONCRETE FLOOR WITH FLOATING SCREED

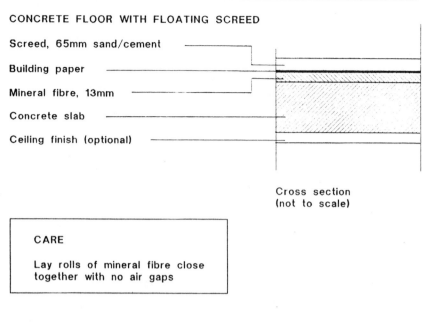

Screed, 65mm sand/cement

Building paper

Mineral fibre, 13mm

Concrete slab

Ceiling finish (optional)

Cross section
(not to scale)

CARE

Lay rolls of mineral fibre close
together with no air gaps

JUNCTION WITH WALL

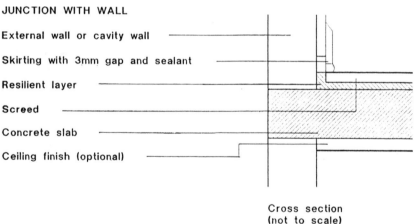

External wall or cavity wall

Skirting with 3mm gap and sealant

Resilient layer

Screed

Concrete slab

Ceiling finish (optional)

Cross section
(not to scale)

*Concrete floors 2.*

## How it is built

The illustrations for *concrete floors 2* and *concrete floors 3* show concrete floor bases with a surface density of 220 kg/m$^2$. The actual thickness of the floor depends on the final density of the concrete that is used. If the concrete has a density of 2400 kg/m$^3$, for example, then the thickness of concrete you need is the fraction 220/2400 of 1 metre, which equals 0.092

CONCRETE FLOOR WITH FLOATING TIMBER RAFT

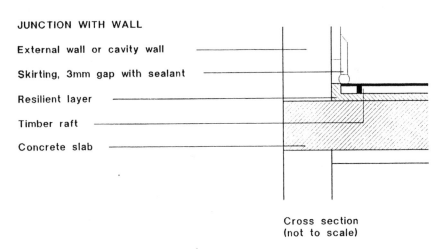

Board floor, tongue & grooved

Timber batten

Mineral fibre, 13mm

Concrete slab

Ceiling finish (optional)

Cross section
(not to scale)

```
CARE

Nails must not penetrate battens
Avoid gaps in the resilient layer
```

JUNCTION WITH WALL

External wall or cavity wall

Skirting, 3mm gap with sealant

Resilient layer

Timber raft

Concrete slab

Cross section
(not to scale)

*Concrete floors 3.*

metres (92 mm). The figure of 220 kg/m$^2$ can include any floor screeds, any shuttering and any ceiling finish bonded to the floor.

The people who actually install a floating floor often find it hard to believe in the total isolation of the two parts of the floor. There seems little harm in just one or two hidden nails, especially where it helps appearance. In fact the whole trouble and expense of the floating floor construction is often wasted by a defective installation, which looks perfect!

## Floating screed

When the floating layer is a screed it should be at least 65 mm thick if it is a cement and sand screed, or at least 40 mm thick if it is a synthetic anhydride screed. The screed should contain 20 to 50 mm of wire mesh to control cracking. The screed should be separated from the flexible layer by a sheet of building paper, which can be laid independently or attached to the resilient layer.

The flexible layer beneath the screed can be of two types. If mineral fibre is used it should have a thickness of at least 13 mm and a density of at least 365 kg/m$^3$. If board material is used it should be pre-compressed expanded polystyrene, rated for impact sound.

## Floating timber raft

The floating layer should be constructed from timber or wood-based boards which are at least 18 mm thick and have tongue and grooves to avoid air paths. The boards are fixed to timber battens at regular intervals but the nails must not penetrate through the battens.

The battens sit on the resilient layer but must *not* be fixed to the floor base beneath. The material for the resilient layer is mineral fibre with a thickness of at least 13 mm and a density of at least 365 kg/m$^3$.

## Junctions

Where a solid floor meets an external wall or a cavity separating wall, the base of the floor, but not the screed, should be passed through the wall. This applies even when the floor beams are running parallel to the wall and the junction is not needed for structural support. If the floor is constructed from prefabricated concrete units such as hollow concrete beams, then the first joint should be at least 300 mm from the face of the wall.

If the floor meets an internal solid wall weighing less than 355 kg/m$^2$ the floor base should be passed through the wall. If the wall is sound resisting or heavier than 355 kg/m$^2$ there is a choice of passing the wall or passing the floor base through.

When the wall is passed through it is important to tie the floor base to the wall and to seal the joint with grout.

For any type of external wall which has openings 20 per cent or less of the external wall area, then the weight of the external wall, of inner leaf of a cavity wall, should have a surface density of at least 120 kg/m$^2$. This figure can include plaster but not dry lining. Where the openings are more than 20 per cent of the external wall area, there are no requirements in the building regulations.

The separation of the two parts of the floor must be maintained all around the edges of the floor. Leave a gap of at least 10 mm between the floating floor and a wall. This gap is filled with mineral fibre, usually by turning up the resilient layer. The skirting board is kept separate from the floating layer by leaving a gap of about 3 mm and filling the gap with sealant.

The overall performance of the floor is also affected by possible flanking sound via construction details on either side of the floor. If a pipe has to penetrate the floor then it should be contained in ductwork made from a board material with a weight of at least 15 kg/m$^2$. All pipework inside the duct should be wrapped in a minimum 25 mm of mineral fibre.

## Variations

*Combinations* of the different types of concrete base and either type of floating layer can be used to give a similar standard of sound insulation.

*Commercial systems* of sound-resistant flooring are available which provide the same level of sound insulation as the constructions described here. Some systems formalise the use of multiple layers of plasterboard and other systems use patent sandwich boards which contain a resilient layer. Such systems work by the same mechanisms of sound insulation given with the examples in this book. They will also fail to work for similar reasons of poor or incorrect installation.

# TIMBER FLOORS WITH FLOATING LAYER

Timber floating floors work in a similar way to the concrete floating floors described previously. The two separate layers of the floor are physically isolated from one another by the use of resilient materials, such as mineral fibre. The top half of the floor just rests on the base of the floor and is held there by the force of gravity. There must be no rigid links between the two parts of the floor, not even a single nail.

A timber floor has less structural weight than a concrete floor so the mass is supplemented by additional layers of board and plasterboard. Airborne sound is also reduced by an absorbent blanket or by *pugging* installed above the ceiling.

When built correctly, the floating floor constructions shown here satisfy the requirements of the Building Regulations for airborne and impact sound. Surveys have found that such standards of sound insulation are regarded as reasonable by most households for most of the time.

### How it works

Refer to the construction details shown in the diagrams for *timber floors 1* and *timber floors 2*.

- The *weight* of the plasterboard in the floating layer provides resistance to airborne sound.
- The acoustical *isolation* of the floating layer provides resistance to airborne sound.
- The acoustical *isolation* of the absorbent blanket on the ceiling provides resistance to airborne sound. Alternatively, the *weight* of the pugging on the ceiling provided resistance to airborne sound.
- The *weight* of the plasterboard in the ceiling provides resistance to airborne sound.
- The floor must be *airtight* in its construction and at its junctions.
- The physical *isolation* of the floating layer gives resistance to impact sound.

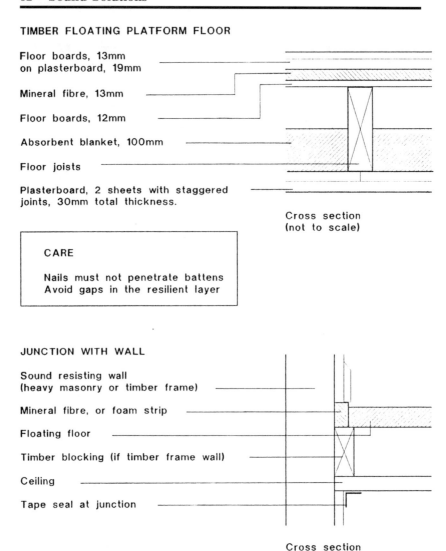

TIMBER FLOATING PLATFORM FLOOR

Floor boards, 13mm
on plasterboard, 19mm

Mineral fibre, 13mm

Floor boards, 12mm

Absorbent blanket, 100mm

Floor joists

Plasterboard, 2 sheets with staggered
joints, 30mm total thickness.

Cross section
(not to scale)

CARE

Nails must not penetrate battens
Avoid gaps in the resilient layer

JUNCTION WITH WALL

Sound resisting wall
(heavy masonry or timber frame)

Mineral fibre, or foam strip

Floating floor

Timber blocking (if timber frame wall)

Ceiling

Tape seal at junction

Cross section
(not to scale)

*Timber floors 1.*

- The isolation must be *complete* and not reduced by rigid links
  or flanking transmissions.

## How it is built

The illustrations for *timber floors 1* and *timber floors 2* show
floor structures built from timber joists in a standard manner but

TIMBER FLOATING RIBBED FLOOR

Board floor, 18mm tongue & grooved
on plasterboard, 19 mm

Timber batten, 50mm width

Mineral fibre, 25mm

Floor joists, 50mm width

Absorbent blanket, 100mm

Plasterboard, 2 sheets with staggered
joints, 30mm total thickness.

Cross section
(not to scale)

```
CARE

Nails must not penetrate battens
Avoid gaps in the resilient layer
```

JUNCTION WITH WALL

Sound resisting wall
(heavy masonry or timber frame)

Mineral fibre, or foam strip

Floating floor

Timber blocking (if timber frame wall)

Ceiling

Tape seal at junction

Cross section
(not to scale)

*Timber floors 2.*

with the layers of floor and ceiling made as thick, and therefore
as heavy, as possible. At some stage there is a layer of springy
material which separates the top of the floor structure from the
bottom.

## Floating platform floor

The floating layer is constructed from timber or wood-based boards which are at least 18 mm thick and have tongue and grooves in order to avoid air paths. These boards are laid on plasterboard which is at least 19 mm thick, or another material with an equivalent weight.

Below the floating layer is a resilient layer of mineral fibre with a minimum thickness of 25 mm and density between 60 and 80 $kg/m^3$. The resilient layer is on a 12 mm floor base of timber or wood-based boards which are nailed to the floor joists.

## Floating ribbed floor

The floating layer is constructed from timber or wood-based boards which are at least 18 mm thick and have tongue and grooves in order to avoid air paths. The boards are fixed to timber battens at regular intervals but the nails must not go through the battens and pierce the mineral fibre.

The battens sit on the floor joists separated by a resilient layer of mineral fibre with a thickness of at least 25 mm and a density between 90 and 140 $kg/m^3$. There must be no rigid links at any point between the floating layer and the floor joists.

## Absorbent ceiling

The ceiling is formed from a 30 mm thickness of plasterboard in two layers. The joints between the plasterboards are staggered so that they are fixed to alternate joists.

A sound absorbent blanket of mineral fibre is laid on top of the ceiling between the joists. The mineral fibre must be at least 100 mm in thickness and have a density not less than 12 $kg/m^3$.

An alternative to the sound absorbent blanket is a layer of dry sand or fine gravel called *pugging* which rests on the ceiling separated by a layer of plastic sheeting. This material must provide a surface density of at least 80 $kg/m^3$, which usually means a thickness of about 50 mm. The purpose of the pugging is to provide extra weight and the joists and the ceiling fixings must therefore be strong enough for the pugging.

## *Junctions*

The separation of the two parts of the floor must be maintained at the edges of the floor. Leave a gap of at least 10 mm between a wall and the floating layer and seal the gap with a strip of mineral fibre or plastic foam. The skirting board is kept separate from the floating layer by leaving a gap of about 3 mm and filling the gap with sealant.

At the junction of the floor with the wall use any form of construction which will block the air paths between floor and cavity. When floor joists are at right angles to the wall, fill the spaces between the joists with timber spacers. Make sure that any gaps in the wall around joists or joist hangers are made good with mortar.

If the wall meets a lightweight masonry wall, with a weight less than 360 kg/m$^2$, then the sound insulation of that wall needs to be upgraded if the *floor* is to achieve a standard of sound insulation similar to the examples above. The sound insulation of the wall is improved by adding an absorbent curtain and plasterboard using a form of construction shown in later sections. Commercial dry lining boards will give a similar effect.

## Variations

An *intermediate platform* in the joist space can be used to support the weight of the pugging and keep it off the ceiling. This platform usually rests on battens attached to the sides of the joists and is a form of construction which may be found in blocks of flats built before concrete floors were in use.

*Commercial systems* of sound-resistant flooring are available and they provide the same level of sound insulation as the constructions described here. Some systems formalise the use of multiple layers of plasterboard and other systems use patent sandwich boards which contain a resilient layer. Such systems work by the same mechanisms of sound insulation given with the examples in this book. They will also fail to work for similar reasons of poor or incorrect installation.

# REMEDIAL INDEPENDENT CEILING

If you live in a flat, you are more likely to be troubled by noise from the people living above you than from people below. Not because sound 'sinks' but because there is the additional problem of impact sound such as footsteps or scraping furniture.

The remedy described here assumes that you do not have access to the room above and therefore concentrates on your ceiling. Or rather, it forgets your existing ceiling and builds a new and independent ceiling. Quieter rooms nearly always mean smaller rooms but, if you are lucky, the older properties which often need remedial sound insulation also have high ceilings and you can afford to sacrifice some space.

The construction of the independent ceiling will improve the sound insulation against both airborne sound and impact sound by between 5 and 10 dB, depending on the frequencies contained in the noise. An improvement of 5 dB is noticeable and an improvement of 10 dB roughly corresponds to a halving in loudness.

Before starting this type of remedial treatment you should try to ensure that most of the troublesome noise is coming through the ceiling and not via flanking routes such as through walls, service ducts or via windows.

## How it works

Refer to the construction details shown in the diagram for *remedial floors*.

- The acoustical *isolation* of the absorbent blanket provides resistance to airborne sound.
- The *weight* of the plasterboard in the ceiling provides resistance to airborne sound.
- The floor must be *airtight* in its construction and at its junctions.
- The physical isolation of the new from the old ceiling structure provides resistance to impact sound.

INDEPENDENT CEILING

Existing ceiling

Minimum spacing, 150mm

Timber joists, 450mm spacing

Mineral fibre, 50mm

Plasterboard, 2 sheets, staggered joints

Cross section
(not to scale)

CARE

Check joist sizes for strength
Seal all joints and junctions

FLOATING PLATFORM FLOOR

New chipboard floor
on plasterboard planks, 19mm

Mineral fibre, 25mm

Existing floor

Floor joists

Absorbent blanket, 100mm

Existing ceiling

Cross section
(not to scale)

*Remedial floors.*

## How it is built

The details of the independent ceiling shown in the *remedial floors* diagram show a new ceiling lining at least 150 mm (6 inches) below the existing ceiling. The new structure must not touch the old ceiling. Seal any holes or cracks in the old ceiling before building the new one.

The new joists should run across the narrowest dimension of the room, providing that those walls can carry the extra load. Timber battens, 50 mm by 50 mm, can be fixed to the wall and the joists notched to as sit on the battens. Plasterboard is heavy and the joists must be deep enough to prevent sagging. Appropriate joist sizes range from 100 mm by 50 mm for a room width of 2.5 metres (8 ft) to 175 mm by 50 mm for a room width of 4.25 metres (14 ft).

All joists are set with their centres separated by 450 mm and this distance will coincide with edges of the standard sizes of plasterboard. Cross braces or noggins are installed, with the help of skew nails, between the joists where plasterboard edges need support. The bottom edges of the battens, joists and noggins must be flush for the plasterboard.

Mineral fibre, such as a quilt of glass fibre or rock wool, is laid over the new joists before the plasterboard is installed. Use two layers of 13 mm plasterboard to line the joists and arrange them so that the joints in the first sheet do not coincide with the joints in the second sheet.

Nail the plasterboards to all joists and noggins at 150 mm intervals using 30 mm long nails for the first layer and 50 mm nails for the second layer. Fill the joints in the plasterboard with appropriate materials and carefully fill in all gaps around the edge of the ceiling.

## Variations

*Joist hangers* can be fixed to the wall although battens will also be needed to fix the edges of the plasterboard.

A *metal lath and plaster* system, of 25 mm thickness, can be used instead of plasterboard.

# REMEDIAL FLOATING FLOOR

This remedy is complementary to the independent ceiling described above. It is assumed that you do not have access to the room below although, by lifting the floorboards, you can get very close to it. The construction of the floating platform on the existing floor gives useful insulation against airborne sounds from below, such as voices and music.

The improvement in insulation is slightly below that given by the independent ceiling. Once again, a quieter room means a smaller room and the floor level will be raised by about 70 mm — almost 3 inches. The skirting boards have to be refitted and there will be problems with details of doors, radiators, sanitary fittings, and floor coverings. It might be easier to move home!

Before starting this type of remedial treatment you should try to ensure that most of the troublesome noise is coming through the ceiling and not via flanking routes such as wall, service ducts or open windows.

## How it works

Refer to the construction details shown in the diagram for *remedial floors*.

- The *weight* of the plasterboard in the floating layer provides resistance to airborne sound.
- The acoustical *isolation* of the floating layer provides resistance to airborne sound.
- The acoustical *isolation* of the absorbent blanket on the ceiling (if installed) provides resistance to airborne sound.
- The floor must be *airtight* in its construction and at its junctions.
- The physical *isolation* of the floating layer gives resistance to impact sound.
- The isolation must be *complete* and not reduced by rigid links or flanking transmissions.

## How it is built

The floating floor shown in the *remedial floors* diagram is a floating platform similar to the ones used for those timber floors which satisfy the Building Regulations for sound insulation. The existing floor joists must be strong enough to withstand the increased weight of the new floor. See the details of the independent ceiling for an idea of suitable joist sizes.

An absorbent blanket is inserted between the joists by either removing the floor boards or by replacing the ceiling. The skirting boards must be removed in either case. For the absorbent blanket between the joists use a 100 mm thickness of mineral fibre (fibre glass or rock wool quilt).

If the floorboards are reasonably sound and plane replace them and use screws in places where they are inclined to squeak. Otherwise nail 12 mm plywood to the joists as the floor base. Lay 25 mm of mineral fibre on the floor base and run it right to the edges of the room.

Plasterboard planks, 19 mm thick, are laid loose on the mineral wool with gaps at the walls. Flooring chipboard, 18 mm with tongue and grooves, is laid on the plaster with all joints glued but not nailed down. A gap of 10 mm is left near the walls.

Use a resilient strip, as shown in the *timber floors* diagrams, to isolate the floor chipboard from the wall. Re-fix the skirting boards so that they cover the gap but do not let them touch the chipboard.

## Variations

If a *new ceiling* is installed then it should have the equivalent mass of 30 mm of plaster. You can use two layers of plasterboard, laid with staggered joints, and with a total thickness of at least 30 mm. Seal all gaps at the edges of the ceiling.

*Commercial systems* are available which use factory-made boards containing dense materials and a resilient layer. If installed correctly they give worthwhile improvements in sound insulation.

# Openings

It is a waste of time building a wall with very high sound insulation unless every part of that wall has the same standard of sound insulation. This ideal can be difficult to achieve because, for all sorts of sensible reasons, many walls have to contains windows, doors and other openings.

# DOUBLE-GLAZED WINDOWS

Windows are usually the weakest link in the sound insulation of a building. The *overall* sound insulation of a wall, for example, is always dragged down to near the insulation value for the windows. A single-glazed window provides about 25 dB of sound insulation and a well-designed double-glazed window can provide over 40 dB of sound insulation. In practice, most double glazing gives about 35 dB of sound insulation.

Double glazing is commonly used to increase the thermal insulation of a window. The layer of air in the cavity and the general improvement in airtightness help prevent the transfer of heat and these features can also help prevent the transfer of airborne sound.

The best airgap for sound insulation has a width of at least 150 mm (6 inches) so that the air does not form a strong linkage between the two sides of the cavity. Unfortunately the best width for thermal insulation is only 20 mm (0.8 inch) because the aim is to prevent the air moving about in the cavity and transferring heat by convection. However, glass is a relatively heavy material and adding an extra pane of glass, anywhere, improves the sound insulation.

Windows are a traditional and logical source of fresh air but ventilation and sound insulation are not compatible in a window. When windows are being designed for sound insulation you must also design special ventilators.

### How it works

Refer to the construction details shown in the illustration for *openings*.

- The *weight* of the glass provides resistance to airborne sound.
- The physical *isolation* of the two frames provides resistance to airborne sound.
- The *width* of the air space provides resistance to airborne sound.
- The acoustical *isolation* of the absorbent material around the

## DOUBLE-GLAZED WINDOWS

Glass, heaviest possible panes
different thicknesses

Airgap, at least 150 mm

Absorbent lining around edge

Resilient glazing gaskets

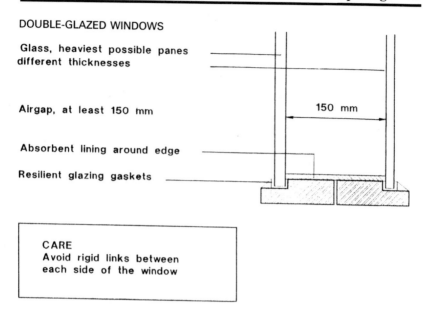

150 mm

CARE
Avoid rigid links between
each side of the window

## AIR VENT

Absorbent lining

Air path with multiple turns

Air inlet

Air outlet

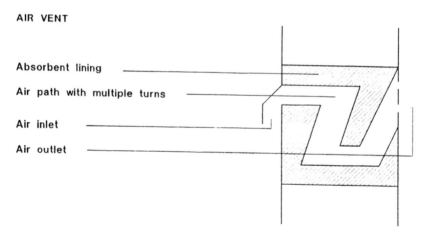

*Openings.*

edges of the air space provides resistance to airborne sound
and prevents resonances between surfaces.

- The *discontinuous construction* of the frames gives resistance
  to flanking sounds around the edges of the window.
- The window must be *airtight* in its construction and at its
  junctions.

## How it is built

To get maximum sound insulation you must maintain the isolation between the two panes. The two sides of the window need to be built in separate frames with no rigid contact between the two sides. Window sills and boards that might normally form a link between the frames must be dislocated at some stage.

The glass forming each pane should be as thick as possible and also, if possible, the two panes should be of different thicknesses in order to discourage resonances. Similarly it helps to have one pane set at a different angle to the other pane, although normally this technique is only used for specialised environments such as broadcasting studios.

### *Airgaps*

The panes of glass should be set into the frames with a compression seal of rubber or neoprene. A flexible mastic sealant should be used to maintain airtightness between the window frame and the wall. The sides of the airgap (the *reveals*) should be lined with an absorbent material like felt.

Like all windows, the insulation is wasted if openable windows are not well-sealed when they are closed. You should aim for the same standard of seal that you would use to keep out draughts, or a ship from sinking! The moment that a window is opened 'just a little' the sound insulation of the entire wall falls to about 10 dB.

## Variations

*Plastic sheeting* is sometimes used for glazing, especially where resistance to damage is important. Plastic sheeting is less dense than glass so the sound insulation value of most plastic systems is a few decibels less than glass.

*Gas filling* is used to increase the thermal insulation of some double glazing units and the acoustic behaviour of argon is similar to dry air. The use of sulphur hexafluoride gas in cavities can increase the sound insulation at speech frequencies but may decrease the insulation against lower frequency sounds such as road noise.

# DOORS

Doors, like windows, are often the cause of poor sound insulation. The *overall* sound insulation of a wall, for example, is always dragged down to near the insulation value for the door which may be as low as 20 dB. For example, an internal wall in a house or flat may have reasonable sound insulation, especially if the wall is brick or block, but this insulation appears low because of the door in the wall.

The most common defect is that a door is not airtight. Compressible seals need to be installed around the edge of the door or the door stops. The type of rubber or expanded plastic strip used for draughtproofing is usually adequate along with a solid draught excluder at the bottom of the door. An air path through a letterbox, even through a keyhole, can be significant and they should have covers.

### How it works

- The door must be *airtight* in its construction and around its edges.
- The *weight* of the door panel provides resistance to airborne sound.

# VENTILATORS

Air paths are needed for fresh air but, unfortunately, they are also a major cause of poor sound insulation. Remember that when a window is only slightly opened the sound insulation is ruined. For most dwellings the various compromises needed to balance fresh air, temperature and noise intrusion can be lived with.

If, however, good acoustic double glazing has been installed to insulate against the noise of an airport or road then the windows must be left sealed and alternative methods of ventilation used. Commercial buildings will usually provide ventilation by means of air conditioning plant but smaller offices and dwellings need to install ventilators with a reasonable sound insulation.

### How it works

Refer to the construction details shown in the illustration for *openings*.

- The *length* of the air path provides resistance to airborne sound.
- The *turns* in the air path provide resistance to airborne sound.
- The sound *absorption* of the lining material provides resistance to airborne sound.
- The ventilator must be *airtight* around its junctions with the surrounding wall.

### How it is built

The exact details of construction depend on the amount of space available for the grille of ventilator and the standard of sound insulation required. The example shown uses the technique of absorbing sound when it is reflected off a surface. A single absorption does not reduce the sound level by much so the aim is to force the sound to have multiple absorptions as it is reflected. The most effective sound insulation path is long and with several turns.

# Vibrations

You may be quite tolerant towards the distant sound of your own water tank filling but you will be less tolerant of hearing the neighbours' plumbing. Noise from boilers, washers and similar equipment can cause annoyance both within their own environment and in nearby dwellings.

Most of this noise is transmitted by structure-borne vibrations and the methods of treating this noise may also apply to noise from washing machines, buzzers, TV and hifi, especially if they are attached to a wall or floor. Before taking action against this type of noise you should try to confirm the source and the type of the sound. It may be that the most annoying components are airborne and flanking sounds, such as through open windows, in which case the noise solutions given in previous sections are also relevant.

# PIPES

A simple treatment of noisy pipes in most homes and many offices is to box-in the pipes. Fortunately this type of treatment is also suitable for hiding the pipes and making decoration easier. Here is a chance to combine sound insulation with thermal insulation and good looks.

## How it works

Refer to the construction details shown in the illustration for *vibrations*.

- The *weight* of the material used to enclose the pipe provides resistance to airborne sound.
- The *airtightness* of the pipe enclosure provides resistance to airborne sound.
- The physical *isolation* between the pipe and the casing provides resistance to structure-borne sound.

## How it is built

Before the pipes are boxed and hidden it is important to isolate them from the building. There should be no rigid links between the pipes and the wall, floor or with the boxing. This is an area where people are tempted to make short cuts which can then be hidden from you. The resilient mineral fibre which will be packed around the pipe can be used to give the required isolation.

If necessary, cut extra clearances in openings so that the pipes can be surrounded by the resilient quilt and therefore don't touch the structure. Where the pipes have to be attached to the structure, use rubber washers under the brackets and grommets around the screws to keep the wall or floor isolated. Or devise a securing strap which wraps around the resilient quilt: see the Product File for ideas.

Almost any type of fibre glass or rock wool insulation can be used as the material which is packed around the pipes. The

BOXED PIPEWORK

Ceiling/floor

Wall

Pipe

Mineral fibre packing

Boxwork, dense board

Cross section
(not to scale)

CARE

Seal all joints and junctions

MACHINE ENCLOSURE

Outer covering

Absorbent lining

Sealing

Flexible coupling

Isolation pads

Machine

Cross section
(not to scale)

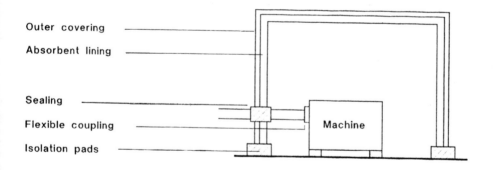

*Vibrations.*

main requirement is that the material is resilient (springy) and has open pores. Although this material will also be good for thermal insulation remember that it is only one part of the sound insulation; the box itself is equally important.

Build the box from the densest board that can be used, rather than from thin plywood or metal. Chipboard, fibreboard and melamine-faced boards are suitably heavy. Finally, make the box airtight by filling gaps with plaster or similar filler.

# FANS AND DUCTS

It is very difficult to pass air without making a noise! If you need to exhaust air from a kitchen of bathroom then you must expect the extractor fan or other system to make some noise. It is a necessary sound as this type of extraction is usually important for ventilation and for the prevention of condensation. So if the operating noise is considered annoying then people are reluctant to switch on the fan, or may even disconnect it.

The noise from fans and air ducts can be made worse by the way they are mounted and connected to one another. For example, if a bathroom extractor fan is simply mounted in a wall or ceiling panel its noise may be amplified. If a duct is connected to such a fan then the noise can also be transmitted to other places.

The following are some general principles for minimising the sound of simple exhaust systems used in homes and offices.

## Fans

- Large slow fans are quieter than small fast fans moving air at the same rate.
- Solid walls provide the best support for fans.
- Panels of glass or other sheet material supporting fans should have as small an area as possible.
- Panels supporting fans should be braced and stiffened.

## Ducts

- Smooth air flows cause the least noise.
- Material for the walls should be as heavy and as stiff as possible.
- The entrance to the fan should be clear of obstacles and bends for at least one fan diameter.
- Struts and obstruction ducts should be avoided.
- Tapers should be used between ducts of different sizes.
- Flexible connectors should be used to link duct and fan.

# MACHINERY

The sound insulation of mechanical plant such as lifts, large boilers, pumps and generators generally deserves specialist construction methods, such as the examples shown in the Product File at the end of the book.

However, the general principles used for the sound insulation of such machinery can also be applied to equipment which we use in homes and offices. The following items are often sources of noise which is transmitted by vibration:

- Central-heating boilers and pumps.
- Phones, entry-phones, bells, buzzers.
- Typewriters and printers.
- Washing machines, dishwashers.
- Television, hifi speakers.

If a typewriter or computer printer, for example, is placed on a table some components of the sound are transmitted into the table which can also act as a sounding board and amplify the sound. This effect is easily reduced by placing the item on a thick pad of rubber or other resilient material.

This type of solution can also be effective for the structure-borne sound from televisions and loudspeakers placed on shelves attached to party walls or built into the chimney recesses. The noise of a washing machine or dishwasher is often transmitted through floors and probably has to be lived with as it is not practical to sit them on resilient mounts. Their noise transmission is usually minimised by having them sitting firm and level on the floor.

Many complaints between neighbours arise from mechanical devices attached to party walls: phones, buzzers, even clocks. The vibrations are transmitted into the wall which may then act as a sounding board, like the table described earlier. This structure-borne transmission can be prevented by having resilient mounting pads between the device and the wall. Securing screws or other fixings also need to be isolated from the device by grommets or by using patent fittings.

It is recommended that some hifi loudspeakers systems are spiked to the floor. The spikes may give better sound but it will certainly increase the noise in the room below, especially in a wooden structure. This type of problem usually calls for negotiation with the neighbours, or for that ultimate in noise control measures: moving house!

# Part Three
# Technical Reference

This part of the book contains descriptions, formulae, figures, regulations and other information which support the earlier parts of the book. The information runs in a similar order to Part One of the book.

Not everyone will need this more technical information and if you are new to the subject then please start with Part One of the book, which gives a 'friendly' account of sound and its effects. Alternatively, you can jump straight in to one of the practical problems and solutions given in Part Two of the book.

# Nature of Sound

## Sound sources

Sound, from a human viewpoint, is a sensation in the brain which is caused by small pressure variations in the air. Changes in the weather or in altitude produce changes in air pressure but the ear does not usually respond to these. Occasionally we 'fool' the ear by rapidly changing altitude, such as when we are in an aircraft.

The pressure changes that cause sound are created when air is in contact with a vibrating source which might be a solid object, a liquid or a gas. Common causes of sound include the following examples:

- Moving cones of paper in loudspeakers.
- Strings on a musical instrument such as a guitar, violin or piano.
- The vocal chords in the human throat.
- Vibrating walls and floors in buildings.
- Vibrating air inside musical wind instruments such as horns and organ pipes.
- Turbulent air created by ventilation fans and jet engines.
- Turbulent liquids created by movement in pipes or bowls.

## Sound vibrations

A vibrating object first produces a *compression* in the adjacent layer of air and then, as it moves away from the air, it leaves a shortage or *rarefaction*. Meanwhile the first compression has been passed outwards and the cycle repeats.

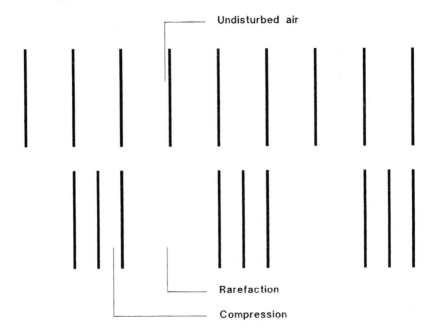

*Vibrations in a sound wave.*

Although it may seem a little untechnical, this view of sound vibrations can be verified by experiments and is closer to reality than the drawings of wave motion shown later.

## Wave motion

Among the properties of sound is the ability to transfer energy from one place to another. From someone's throat to someone else's ear, for example. Although the particles of the intermediate material, the air, vibrate, they do *not* change their basic positions. The energy associated with sound is transferred by a *wave motion* where the 'front' of the wave spreads out equally in all directions unless it is affected by an object or another material in its path.

Sound waves are called *longitudinal* waves because the particles of the medium carrying the wave, such as air or brick, vibrate in the same direction as the direction of the wave. The other type of wave is a *transverse* wave, such as those in the sea, where the particles of the medium vibrate at right angles to the direction of travel.

Amplitude

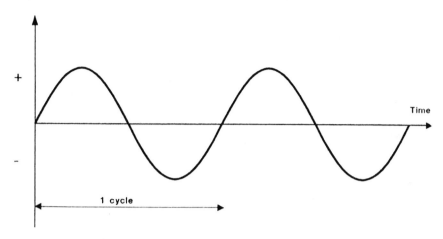

*Model of sound wave.*

It is difficult to depict a longitudinal wave in a diagram so we usually show its vibrations in a different dimension, which makes it look like a transverse wave. This is a convenient model which gives us useful results; but it also helpful to remember that the vibrations of sound are actually occurring back and forth along the line of travel, as shown in the earlier diagram.

### Wavelength, frequency, velocity

All wave motions can be described in terms of wavelength, frequency and velocity which are defined as follows.

- *Wavelength* ($[\lambda]$) is the distance between any two repeating points on a wave.
  UNIT: metre (m)
- *Frequency* ($f$) is the number of vibrations per second.
  UNIT: hertz (Hz)
- *Velocity* ($v$) is the distance moved per second in a particular direction.
  UNIT: metres/second (m/s)

For every vibration of the sound source the waves moves forward by one wavelength. The number of vibrations per second therefore gives the total length moved in 1 second, which is another way of expressing velocity. This relationship can be written as the following formula.

$$v = f \times L$$

Where:

$v$ = velocity in m/s
$f$ = frequency in Hz
$L$ = wavelength in m

## Velocity of sound

The 'front' of a sound wave travels outwards from its source with a steady velocity that is *independent* of the frequency of vibrations. In other words, high-pitched notes travel at the same speed as low-pitched notes. In air at 20°C the velocity of sound is 340 m/s, which is slow enough to be noticeable in large buildings.

The velocity of sound *is* affected by the properties of the material that it travels through. The velocity of sound in air increases as the temperature or the humidity increases but it is unaffected by variations in atmospheric pressure such as those caused by the weather.

Sound travels faster in liquids and solids than it does in air because the densities and elasticities of those materials are greater than air. Surprisingly perhaps, this means that sound travels about 10 times faster through a solid wall than through the air, although we are hardly likely to notice the effect.

It can be shown that the velocity of sound in a material is given by the following formula.

$$v = \frac{E}{\rho}$$

Where:

$v$ = velocity of sound in m/s
$E$ = modulus of elasticity in N/m$^2$
$\rho$ = density of material in kg/m$^3$.

## Frequency of sound

If an object or surface vibrates 100 times per second, for example, then the frequency of the sound waves produced will be 100 Hz. Human hearing interprets frequency as *pitch*, which is heard as 'low' or 'high' notes.

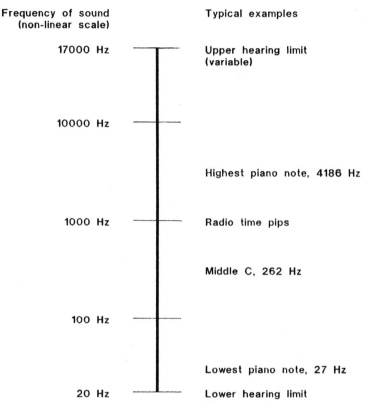

| Frequency of sound (non-linear scale) | Typical examples |
|---|---|
| 17000 Hz | Upper hearing limit (variable) |
| 10000 Hz | |
| | Highest piano note, 4186 Hz |
| 1000 Hz | Radio time pips |
| | Middle C, 262 Hz |
| 100 Hz | |
| | Lowest piano note, 27 Hz |
| 20 Hz | Lower hearing limit |

*Frequency of sound waves.*

The human hearing system responds to frequencies in the approximate range from 20 Hz to 20000 Hz. On a piano, for example, the lowest note is 27 Hz; middle C is 262 Hz and the highest note is 4186 Hz. To help analyse the range of frequencies we usually consider them in ranges of frequency.

## Octaves

- An *octave band* is the range of frequencies between any one frequency and a doubling of that frequency.

For example, 2000 Hz is exactly one octave above 1000 Hz, or 880 Hz is exactly one octave above 440 Hz. An octave interval is significant in music because it sounds very harmonious but the octave is also used in technical work.

A pure tone containing only one frequency can be made by a tuning fork or electronic signal generator but it is rarely found in everyday noises. Most practical sounds contain a combination of many different frequencies. If a single *fundamental* frequency is dominant then we recognise a particular note.

### Overtones and harmonics

- *Overtones* or *harmonics* are frequencies equal to whole number multiples of the fundamental frequency.

The distinctive quality or *timbre* of different voices and instruments depends on the different mixtures of overtones which accompany the fundamental. The frequencies of these overtones may well rise above 10 000 Hz and their presence give the 'hi' of hifi. A domestic telephone line for example doesn't usually pass frequencies much above 3000 Hz and the exclusion of the higher overtones noticeably affects the quality of the sound.

### Resonance

Every object has a *natural frequency* at which it will vibrate when disturbed. For example, you can hear the difference between a metal bar or a block of wood when you drop them on the floor.

- *Resonance* occurs when the natural frequency of an object coincides with the frequency of any vibrations applied to the object.

### Frequency, wavelength calculation

If the velocity of sound is taken as 340 m/s and the frequency is given, then the formula $v = f \times L$ can be used to find the wavelength.

For example, to calculate the wavelengths of the sounds at the extreme ends of the human hearing range which have approximate frequencies of 20 Hz and 20 000 Hz:

For 20 Hz

$$v = f \times L$$
$$340 = 20 \times L$$
$$L = \frac{340}{20} = 17 \text{ metres}$$

For 20 000 Hz

$$v = f \times L$$
$$340 = 20\,000 \times L$$
$$L = \frac{340}{20\,000} = 0.017 \text{ metres or 17 mm}$$

To be significantly deflected or absorbed, sound waves need to meet physical obstacle or texture of similar dimensions to the wavelength. The above answers help explain why it is easier to absorb high frequency sounds than low frequency sounds: the high frequency sounds have waves with small wavelengths which make them more liable to deflections and absorption.

## Strength of sound

The strength of a sound depends upon its energy content which, in turn, affects the size of the pressure variations. The amplitude of the sound wave, the maximum displacement of each air particle, is the property that we perceive as loudness, a term which is discussed later.

To measure the strength of a sound it is necessary to measure some aspect of its energy or pressure, such as the following properties.

- *Sound power* (*P*) is the rate at which sound energy is produced at the source.
  UNIT: watt (W)
- *Sound intensity* (*I*) is the sound power measured over the area upon which it is received.
  UNIT: watts per square metre (W/m$^2$)

Amplitude

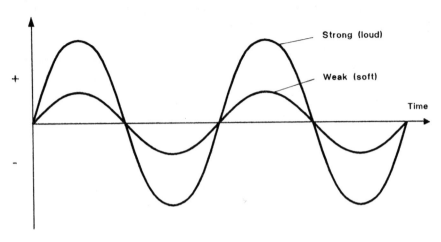

*Strength of sound waves.*

- *Sound pressure* (*p*) is the RMS average variation in pressure caused by the sound.
  UNIT: pascal (Pa) where 1 Pa is the same as 1 N/m$^2$

The pressure of sound waves continuously varies between positive and negative values so it is measured as a *root mean square* (RMS) value which gives positive values only.

### Intensity−pressure formula

The intensity of a sound is proportional to the square of the pressure given by the following formula.

$$I = \frac{p^2}{\rho \, v}$$

Where:

$I$ = intensity of the sound (W/m$^2$)
$p$ = pressure of the sound (Pa)
$\rho$ = density of the material (kg/m$^3$)
$v$ = velocity of sound (m/s).

For air the value of $\rho$ v can be taken as 410.

## Thresholds of hearing

- The *threshold of hearing* is the weakest sound that average human hearing can detect.

The value of the threshold varies slightly with individual hearing but is remarkably low and occurs when the membrane in the ear is deflected by a distance less than the diameter of a single atom. For reference purposes the threshold of hearing is defined to have the following values at 1000 Hz.

$I = 1 \times 10^{12}$ W/m$^2$ when measured as intensity.
$p = 20 \times 10^{-6}$ Pa (or 20 micropascals) when measured as a pressure.

- The *threshold of pain* is the strongest sound that the human ear can tolerate.

This high threshold is less distinct than the threshold of hearing because the effect of high pressures on the ear gradually changes from sound to pain and then to physical damage. The threshold has the following approximate values.

$I = 1$ W/m$^2$ when measured as intensity.
$p = 200$ Pa when measured as a pressure.

## Decibel scale

Absolute measurements of intensity or pressure, as described above, do accurately describe the strength of a sound wave but they give inconvenient numbers and do not correspond well to the way that the hearing system judges changes in sound strength.

For practical measurements of sound strength we therefore usually use a decibel scale based on a logarithmic ratio. Measurements made in electronics also use decibel values but they should not be confused with those used for sound level.

Decibel values are calculated by the formulae:

$$N = 10 \log_{10} (I_2/I_1)$$

or

$$N = 20 \log_{10} (p_2/p_1)$$

Where

$N$ = number of decibels
$I_1$ and $I_2$ are the two intensities being compared.
$p_1$ and $p_2$ are the two pressures being compared.

## Sound levels

The decibel measurement of sound is formed by the ratio between the measured sound and the threshold of hearing. The term 'sound level' implies that the standard value of the threshold of hearing is used, as shown in the numerical example.

Converting absolute values of energy or pressure to sound levels in decibels produces a scale of numbers that is convenient and gives a reasonable correspondence with the way that the ear compares sound, as shown in the diagram.

The smallest change that the ear can detect is about 1 dB, but a 3 dB change is the smallest difference that is normally considered significant. A 3 dB change corresponds to a doubling (or halving) of sound energy but it does *not* sound twice as loud. A 10 dB increase or decrease makes a sound seem approximately twice as loud or half as loud.

## Calculation of sound levels

Values of sound intensity or sound pressure are converted to decibels by comparing them with the threshold of hearing, on a logarithmic basis.

- *Sound intensity level* (SIL) is used to calculate the decibel value when sound strength is measured in terms of intensity.

$$\text{SIL} = 10 \log \frac{I}{I_0}$$

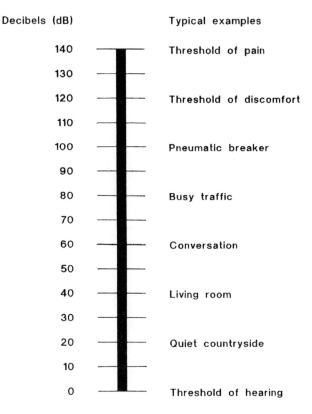

| Decibels (dB) | | Typical examples |
|---|---|---|
| 140 | | Threshold of pain |
| 130 | | |
| 120 | | Threshold of discomfort |
| 110 | | |
| 100 | | Pneumatic breaker |
| 90 | | |
| 80 | | Busy traffic |
| 70 | | |
| 60 | | Conversation |
| 50 | | |
| 40 | | Living room |
| 30 | | |
| 20 | | Quiet countryside |
| 10 | | |
| 0 | | Threshold of hearing |

*Decibel scale of sound level.*

Where:

$I$ = intensity of the sound being received (W/m$^2$)
$I_0$ = intensity of the threshold of hearing taken as $1 \times 10^{-12}$ W/m$^2$.

- *Sound Pressure Level* (SPL) is used to calculate the decibel value when sound strength is measured in terms of pressure.

$$\text{SPL} = 20 \log \frac{p}{p_0}$$

Where:

$p$ = RMS pressure of the sound being received (Pa)
$I_0$ = RMS pressure of the threshold of hearing taken as 20 µPa ($20 \times 10^{-6}$ Pa).

Most instruments measure sound by responding to the sound pressure and, for practical purposes, the SIL and SPL give the same value in decibels.

### Sound level calculation

If a certain sound has an average pressure of $4.5 \times 10^{-2}$ Pa then the appropriate formula can be used to calculate the sound pressure level (in decibels). The standard value for the threshold of hearing is used and, in terms of pressure, it is $20 \times 10^{-6}$ pascals.

$$\text{SPL} = 20 \log \frac{p}{p_0}$$

Substituting the known values into the formula gives

$$\text{SPL} = 20 \log \frac{4.5 \times 10^{-2}}{20 \times 10^{-6}}$$

The division inside the bracket must be done first. On a calculator, press the Exponent button (often labelled EE) to enter the powers of 10, such as $-1$ and $-6$. Don't enter $\times 10$ as this is taken for granted when you press the Exponent button. The intermediate result is:

$$\text{SPL} = 20 \log(2250)$$

On a calculator press the button for logarithm to base 10 (often labelled log). Don't use the natural logarithm (often labelled ln). The result is:

$$\text{SPL} = 20 \times 3.3522$$
$$= 67.04$$

As human hearing is only accurate to within about 1 decibel the answer can be written as

Sound Pressure Level = 67 dB.

## Combination of sound levels

If several different sounds are heard at the same time then the ear receives extra energy or pressure. Because the decibel scale is constructed using logarithms you cannot simply add the sound levels in decibels. For example, two engines each with an SPL of 100 dB do not combine to give a total SPL of 200 dB; which is just as well as that level of pressure could literally blow you away.

Although decibel values cannot be directly added, you are allowed to add intensity values (W/m$^2$) or the square of pressure values (Pa):

$$I = I_1 + I_2$$

or

$$p = \sqrt{(p_1^2 + p_2^2)}$$

If you experiment with the formulae for sound intensity level you can verify a simple practical rule that

• Doubling the energy gives a 3 dB increase in sound level.

So when you double the intensity (energy) of a sound, such as starting two engines of 100 dB each, then the new sound level is $100 + 3 = 103$ dB.

If the difference between two sounds is greater than 15 dB then the addition of the lower level will produce negligible effect. For example, the combination of a 100 dB sound with an 80 dB sound is $100 + 0 = 100$ dB, so that the 100 dB sound masks the 80 dB. This seems strange and total masking only occurs if the sounds are of similar frequency structure, such as two trains. Sounds like telephone warblers are designed to have a distinctive frequency structure that will be heard in spite of higher surrounding sound levels.

# Hearing

## Human hearing system

Our sense of hearing involves the ear, the brain and the connection between them. The function of the ear is to convert sound waves into nerve impulses and it can be considered in three main parts, as shown in the diagram of the hearing system.

- *The outer ear* is the part of the ear that can be seen. It collects the sound waves, funnels them onto the skin of the eardrum which then vibrates.
- *The middle ear* is an air-filled cavity which passes the vibrations of the eardrum to the inner ear by means of three small bone levers.

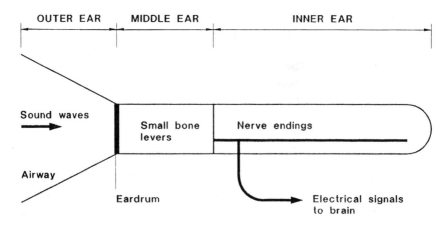

*Hearing system.*

- *The inner ear* is a hollow coil of bone filled with liquid and containing thousands of fine hairs attached to nerve endings. This system converts the sound vibrations to electrical signals which are transmitted to the brain by the auditory nerve.

## Human hearing range

The sensitivity of the hearing system to sound waves depends on the frequency (low pitch to high pitch) of the sound and on the strength of the sound (low sound level to high sound level). Different people have slightly different hearing sensitivities but average values can be measured and they give a 'map' of sounds that the ear can detect.

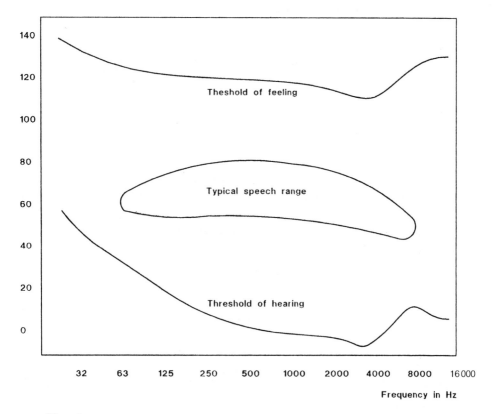

*Hearing range.*

## Hearing loss

Most types of deafness involve a loss of sensitivity over certain ranges of frequency. This loss of hearing is measured by the process of audiometry and has various causes.

*Conductive deafness* is the presence of 'mechanical' faults in the system which conduct sound vibrations to the inner ear. It can be caused by defects such as a broken eardrum, blockage by wax or a stiffening of the bones in the middle ear. There are various medical remedies for this type of deafness.

*Nerve deafness* is a result of damage to the nerve endings of the inner ear or to the nerve which carries information to the brain. This type of deafness can be caused by infections, head injuries and exposure to excessive noise; it is irreversible and is almost impossible to help.

The hearing loss caused by noise exposure is unnecessary and the methods of measuring and predicting noise exposure are discussed in a later section. There are two main effects on the hearing system.

- *Temporary threshold shift* (*TTS*) is a temporary loss of hearing which recovers in one to two days after the exposure to excessive noise
- *Permanent threshold shift* (*PTS*) is a permanent loss of hearing caused by longer exposure to excessive noise.

The way that these hearing losses work can, unfortunately, increase the likelihood of deafness. For example, when people are exposed to excessive noise in a factory or a disco they will initially find the excessive noise uncomfortable and suffer temporary threshold shift. They could recover from this TTS but if they return to the same environment within 24 hours then the noise will not seem as objectionable because their hearing is less sensitive.

Although the noise will now seem more acceptable, the energy level of the noise remains just as damaging and the temporary threshold shift changes into permanent threshold shift. Noisy factories are becoming less common but you are just as likely to suffer PTS by listening to music at high levels through the headphones of your personal stereo system.

Like most forms of deafness, PTS is a creeping effect which

degrades certain regions of hearing sensitivity before others. Initially there is loss of hearing in the high tones around 4000 Hz which may remain unnoticed except perhaps for the misunderstanding of certain sibilant words.

*Presbyacusis* is a gradual loss of hearing sensitivity which accumulates to everyone as they get older. The higher frequencies are affected first but for most people the effect will not be noticed until after the age of 65. However, loss of hearing caused by age adds to any loss of hearing caused by noise exposure earlier in life and the combination of the two effects is usually troublesome.

## Loudness

Although the word 'loud' is common in normal speech it also has a stricter technical meaning. The sensation of loudness in human hearing effect is a function of reception by the ear and interpretation by the brain. Loudness of a particular sound depends on the amplitude (strength) of the sound wave and also on the frequency (pitch) of the sound.

Human hearing is not equally sensitive at all frequencies, as is seen from a map of hearing sensitivity. Therefore sounds of a different pitch (highness or lowness) will be judged to be of different loudness, even when they have the same sound level in decibels as shown on a sound meter.

For example, a 50 Hz tone must be boosted 15 dB so as to sound equally loud as a 1000 Hz tone. This can be demonstrated by asking someone to equalise, by ear, the volume controls for two loudspeakers while one of the speakers is playing a bass note and the other one playing a treble note. If the results are checked with a sound meter or other equipment it is found that everyone turns up the setting of the bass control to get the same 'loudness' as the treble control.

The results of many measurements of human hearing response can be presented in the form of standard contours, as shown in the diagram. It can be seen that the ear is most sensitive in the frequency range between 2 kHz and 5 kHz, and is least sensitive at low frequencies or at extremely high frequencies. When this effect is measured at different sound levels (decibels) then the shape of the response effect changes slightly to give a 'family' of curves.

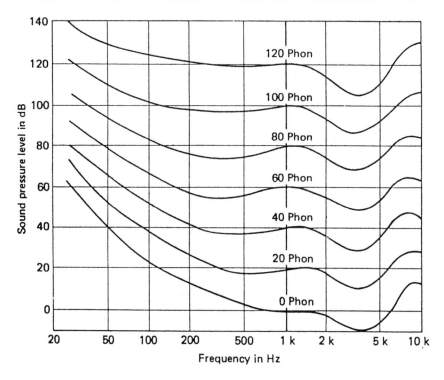

*Equal loudness contours.*

## Loudness level

If two different tones seem equally loud then it can be useful to have a scale which gives them the same value, even though the two tones have different sound levels. The *phon scale* of loudness level is obtained from the family of equal loudness curves.

Pure tones of different frequencies are compared to a reference tone of 1000 Hz. When a tone is judged to be equally loud as the reference tone its loudness level (in phons) is equal to the sound level (in decibels) of the reference tone.

For example: For a tone of 1000 Hz, 60 phons is exactly equal to 60 dB; because 1000 Hz is the arbitrary reference frequency. But the equal loudness contour for a tone of 50 Hz shows that 60 phons is equal to 78 dB. Both sounds would seem equally loud but the lower-pitched tone needs a higher sound level because our hearing system is not as sensitive at low frequencies.

The *sone scale* of loudness is a re-numbering of the phon scale so the sone values are directly proportional to the magnitude of the loudness. For example, 2 sones is twice as loud as 1 sone. One sone is equivalent to 40 phons. Loudness, in sones, is doubled each time that the loudness level is increased by 10 phons.

# Noise

There are a variety of methods and scales in use for assessing noise, depending on the type of situation. The measurement of the noise usually includes the sound levels, the frequencies present, and the duration of the noise. This set of measurements is usually processed to give a convenient single number or index but it should be remembered that the contents of the single number can be complex.

All assessments of noise need to be compared to human opinion about the annoyance or acceptability of different levels of a particular noise. Despite the technicalities of noise measurement, the purpose of it all is to ensure our own comfort and protection.

## Sound meter

Measurements of noise usually start with a sound level meter which converts the variations in air pressure to a variation in voltage. These electrical signals are amplified and displayed on a scale which has been calibrated against known sound levels.

## A-scale dB(A)

A *weighting network* is an electronic circuit in a sound level meter which gives weight to the same middle frequencies that

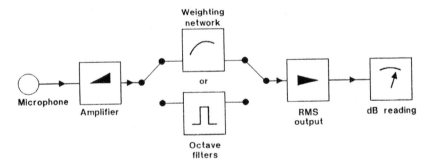

*Components of sound level meter.*

human hearing emphasises. Standard electronic weighting are published and the A-scale, giving results in dB(A), is found to be the most useful.

## Frequency spectra

A sound level measurement that combines all frequencies into a single reading, such as dB(A), is convenient but does not reveal details of the frequencies present. It is more useful to take separate sound level readings with electronic filters which measure over standard ranges of frequency such as octaves or one-third octaves.

The series of measurements can be presented as a plot of sound level against frequency.

## Noise limiting curves

The frequency spectrum of a noise at different frequencies can be compared to standard curves of noise which are based on the

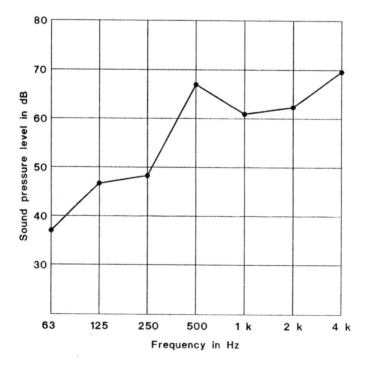

*Noise spectrum.*

sensitivity of the human ear. A single-figure index can be read from the curves.

*Noise criteria (NC)* and *Noise rating (NR)* are used to assess the background noise made by machinery such as heating and ventilating equipment.

## Traffic noise level $L_{10}$

The nuisance caused by traffic depends on the way that the noise varies during a standard period of 18 hours and the $L_{10}$ index takes these variations into account. The $L_{10}$ index is a statistical type of measurement which is best measured by a special sound level meter using standard methods.

A typical measurement at ten metres from the edge of a motorway is $L_{10}$ (18 hours) = 75 dB(A).

## Equivalent continuous sound level $L_{eq}$

Exposure to occupational noise or environmental noise depends on both the sound levels of the noise and the time for which it lasts. The nuisance level of a noise and the risk of hearing damage is found to depend on the total noise energy received in a given period.

- $L_{eq}$ is that continuous sound level which gives the same total energy as the varying sound level.

In a simple example, a doubling of sound energy increases the sound level by 3 dB. Therefore a sound level of 90 dB(A) for a continuous eight hours and a sound level of 93 dB(A) for four hours (followed by four hours of relative quiet) both give the same noise 'dose' written as $L_{eq}$ (8 hour) = 90 dB(A).

An index of $L_{eq}$ (8 hour) = 90 dB(A) is a currently accepted limit for exposure to industrial noise. People vary in their susceptibility to hearing damage and the noise should be below this limit wherever possible.

An index of $L_{eq}$ (12 hour) = 75 dB(A) has been used as an acceptable level for noise nuisance measured at the edge of a building site.

## Noise and number index (NNI)

The nuisance caused by noise near an airport depends on both the noise level of aircraft and the number of take-offs and landings. The NNI is a single index which combines the perceived noise level (PNdB) of aircraft engines and the number of movements over a specified period.

# Sound Effects

## Attenuation

When sound waves spread out from a source they die away or *attenuate*. For the initial prediction of sound attenuation it is simplest to assume a single source of sound which acts like a point and from which the sound waves spread into a *free field* where there are objects.

The results of the pure theory usually need to be modified for practical conditions where the source of sound is often directional, like a jet engine, and the sound waves are reflected and absorbed by objects in their path. The exact effects depend on the relative sizes of objects and the wavelength of the sound.

In order to significantly affect a sound wave, an obstructing object must be similar in size to one wavelength of the sound wave. For sounds with frequencies in the range of the human voice, say 250 Hz, the wavelengths are around one metre in length.

## Inverse square law

- The sound intensity from a point source of sound decreases in inverse proportion to the square of the distance from the source.

This relationship can also be expressed in the following equation.

$$I = \frac{1}{d^2}$$

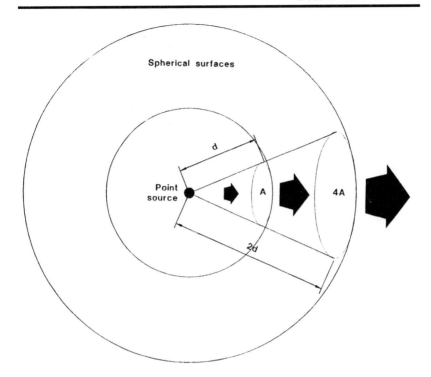

*Spread of sound from point source.*

Where:

$I$ = intensity of sound measured at distance $d$ from the source.

The ratio of any two intensities is given by the formula

$$\frac{I_1}{I_2} = \frac{d_2^2}{d_1^2}$$

Where:

$I_1$ = sound intensity measured at distance $d_1$ from the source.
$I_2$ = sound intensity measured at distance $d_2$ from the source.

The change in sound levels is given by the following formula.

$$\text{Change in SPL} = 10 \log \frac{d_2^2}{d_1^2}$$

UNIT: decibel (dB)

## *Sound and distance calculation*

A microphone measures sound at a position in a free field five metres from a point source. Calculate the change in sound level if the microphone is moved to a position ten metres from the source.

Let $d_1$ = distance 5 m, and $d_2$ = distance 10 m. Substituting in the formula

$$\text{Change in SPL} = 10 \log \frac{d_2{}^2}{d_1{}^2}$$

$$= 10 \log \frac{10^2}{5^2}$$

$$= 10 \log \frac{100}{25}$$

$$= 10 \log 4 = 10 \times 0.6021$$

$$= 6.021$$

$$\text{Change in SPL} = 6 \text{ dB.}$$

Notice that this result agrees with the theory that doubling the distance from a point source of sound in free space changes the sound level by 6 dB.

## Attenuation effects

The passage of sound in the open air is subject to other effects in addition to the effect of distance.

*Air absorption* occurs because the sound wave uses some energy in moving air molecules. The effect is negligible at low frequencies and at 2000 Hz causes a reduction of about 0.01 dB per metre of travel.

*Temperature gradients* are naturally occurring changes in air temperature which cause the path of sound waves to be deflected. One general result of this refraction is that sound travels along the ground better at night than during the day.

*Wind* can deflect the path of sound waves by physically moving the air molecules through which the sound waves are passing.

*Ground effects* can cause some sound energy to be absorbed as sound waves pass over the surface of the ground.

## Sound absorption

Sound absorption is a reduction in the sound energy reflected by the surfaces of a room and has a major effect on room acoustics, especially on reverberation times.

*Absorption coefficient* (*a*) is a measure of the amount of sound absorption given by a particular surface. The amount of sound energy absorbed (not reflected) is compared with the amount of sound energy arriving (incident) at the surface in the following formula.

$$a = \frac{\text{Absorbed sound energy}}{\text{Incident sound energy}}$$

A perfect absorber would have an absorption coefficient of 1.0 and a poor absorber (a good reflector) has an absorption coefficient near zero. A surface that absorbs 40 per cent of incident sound energy has an absorption coefficient of 0.4.

Different materials and construction have different coefficients and the coefficients for any one material changes with the frequency of sound.

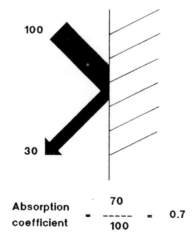

$$\text{Absorption coefficient} = \frac{70}{100} = 0.7$$

*Sound absorption.*

**Table of absorption coefficients**

| Common building materials | Absorption coefficient | | |
|---|---|---|---|
| | **125 Hz** | **500 Hz** | **2000 Hz** |
| Brickwork, plain | 0.02 | 0.03 | 0.04 |
| Concrete, plain | 0.02 | 0.06 | 0.05 |
| Carpet, thick pile | 0.10 | 0.50 | 0.60 |
| Curtains, medium weight, folded | 0.10 | 0.40 | 0.50 |
| Floor boards, on joists | 0.15 | 0.10 | 0.10 |
| Glass, 4 mm in window | 0.30 | 0.10 | 0.07 |
| Plasterboard, on studs with airgap | 0.30 | 0.10 | 0.04 |
| Tiles, ceramic on solid backing | 0.01 | 0.01 | 0.02 |
| **Special items** | | | |
| Audience, per person | 0.21 | 0.46 | 0.51 |
| Seats, empty fabric, per seat | 0.12 | 0.28 | 0.28 |

## Types of absorber

Absorbers reduce sound levels by converting the energy of sound waves into mechanical vibrations in the absorbing structure. This mechanical energy is then usually converted to frictional heat energy. The amount of energy involved in even the loudest sound is actually low and there is no danger of fire!

The materials and the construction used especially for the purpose of absorbing sound are classified in three groups which have different effects at different frequencies.

- (1) *Porous absorbers* are effective at high frequencies, such as those above 1000 Hz.
  Porous absorbers consist of cellular materials such as mineral wool used for acoustic blankets and fibre boards used for acoustic tiles.
- (2) *Panel absorbers* are effective at low frequencies, such as those below 500 Hz.
  Panel or 'membrane' absorbers are fixed sheets of material with a space behind them which can contain air or an absorbent material. The panels may be made of materials such as plywood or they may already exist in the form of windows or suspended ceilings.

● (*3*) *Cavity absorbers* are effective over a specific narrow band of lower frequencies.

Cavity absorbers or *Helmholtz resonators* are narrow en-closures of air with one narrow opening. The cavity may contain air or another material and in practice be part of a continuous structure, such as a perforated acoustic tile.

The maximum absorption occurs at the resonant frequency of the cavity which is given by the following formula.

$$f = 55 \sqrt{(a/dV)}$$

Where:

$a$ = cross-sectional area of the opening ($m^2$)
$d$ = length of the neck of the opening ($m$)
$V$ = volume of the cavity ($m^3$).

The ability to tune a cavity absorber to specific frequencies is useful for damping certain sounds inside rooms or concert halls.

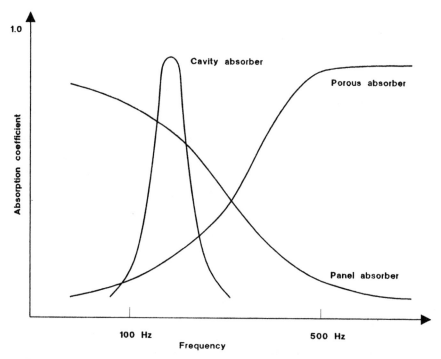

*Response of different absorbers*

## *Practical absorbers*

Acoustic materials such as tiles and panels often absorb sound by using a combination of methods which work best at different frequencies. A fibrous tile material absorbs sound at higher frequencies and it may be drilled with holes which act as small cavity absorbers. Some absorbent materials have a perforated covering to give a resonator effect and if the material is mounted as a panel then it will act as a panel absorber for lower frequencies.

## Reverberation time

With accurate instruments it is difficult to decide just when the reverberant sound has died away, so a drop of one millionth in power (60 dB) is used.

● *Reverberation time* is defined as the time taken for a sound to decay by 60 dB from its original level.

The time taken for this decay depends upon the following factors.

● The distances between the surfaces of the room.
● The absorption at those surfaces.
● The frequency of the sound.

### Table of suitable reverberation times

| Room | Volume ($m^3$) | Reverberation time (s) |
|---|---|---|
| Offices | 30 | 0.5 |
| | 100 | 0.75 |
| Conference rooms | 100 | 0.5 |
| | 1000 | 0.8 |
| Music studios | 500 | 0.9 |
| | 5000 | 1.5 |
| Churches | 500 | 1.5 |
| | 5000 | 1.8 |

## Ideal reverberation times

The optimum reverberation time depends on the volume of the room and the types of sound it is used for.

## Sabine's formula

Sabine's formula is found to give reasonable predictions of reverberation time for rooms without excessive absorption. The formula uses a knowledge of the surface areas of the room and their absorption characteristics to predict the reverberation time for those conditions.

$$RT = \frac{0.16V}{A}$$

Where:

$RT$ = reverberation time (seconds)
$V$ = volume of the room ($m^3$)
$A$ = total absorption of room surfaces ($m^2$ sabins),
   = Sum of (each area × absorption coefficient for that area).

### *Reverberation time calculation*

A lecture hall with a volume of 480 m$^3$ has the surface finishes, surface areas, and absorption coefficients (at 500 Hz) given below.

| | | | |
|---|---|---|---|
| Walls, plaster on brick | 208 $m^2$ | (0.02) | |
| Floor, carpet tiles | 160 $m^2$ | (0.30) | |
| Ceiling, plasterboard | 160 $m^2$ | (0.10) | |

| Surface | Area | Absorption coeff. (500 Hz) | Absorption units |
|---|---|---|---|
| Walls | 208 | 0.02 | 4.16 |
| Floor | 160 | 0.30 | 48 |
| Ceiling | 160 | 0.10 | 16 |
| Occupants | 100 people | 0.46 each | 16 |
| Total A | | | 114.16 |

Calculate the reverberation time at 500 Hz of this hall when it is occupied by 100 people who each contribute 0.46 absorption units.

Tabulate the information and calculate the absorption units for each area using: Absorption = Area × Absorption coefficient.

Using Sabine's formula

$$RT = \frac{0.16 \ V}{A} = \frac{0.16 \times 480}{114.16} = 0.67$$

Reverberation time = 0.67 seconds at 500 Hz.

### Sound reflection

When sound is reflected it obeys the same law of reflection that applies to all waves.

● The angle of reflection equals the angle of incidence.

The angle is always measured between the path of the sound wave and the 'normal' to the curve, where the normal is a line at right angles to the surface.

At curved surfaces, the angle of reflection still equals the angle of incidence and the effect depends on the geometry.

*Concave* surfaces (curving inwards) tend to focus waves and concentrate sound in certain areas. This can lead to strange acoustic results, as in some domed buildings.

*Convex* surfaces (curving outwards) tend to scatter and disperse sound waves, which is generally an acoustically good effect.

*Parallel* smooth surfaces in smaller rooms can set up repeated reflections which give a 'flutter' or 'buzzing' effect to the acoustic quality.

### Echoes

An echo is a long-delayed reflection. Initially a reflected sound reinforces the direct sound, as in reverberation. But if the reflection is delayed and is also strong then this echoes blurs and confuses the original sound.

There is a risk of a distinct echo if a strong reflection is received later than $\frac{1}{20}$th second (50 ms) after the reception of the direct sound. When sound has a typical velocity of 340 m/s, this time difference corresponds to a path difference of 17 m. This difference in length between the direct path and the reflected path can be checked by the geometry of a room and will tend to occur near the front of a hall.

### Sound transmission

Sound passes through a partition, such as a wall, by converting to equivalent mechanical vibrations in the material of the wall. The nature of these vibrations depends upon the way that they are started: by air vibrations or by direct mechanical vibrations. The sound transfer through a partition is classified as either airborne or impact sound, in relation to the receiving room only.

- **Airborne sound** is sound which travels through the air *before* reaching the partition under consideration.
- **Impact sound** is sound which is generated *on* the partition under consideration.

Typical sources of airborne sound include voices, radios, musical instrument, traffic and aircraft noise. Typical sources of impact sound includes footsteps, slammed doors and windows, noisy pipes and vibrating machinery. A continuous vibration can be considered as a series of impacts and impact sound can also be called *structure-borne sound*.

All sound is airborne in the sense that it finally travels through the air to your ear and a classification of airborne and impact sound is only valid for a particular partition considered from the listening side.

A single source of noise may generate both airborne and impact sound, depending on the listener. For example, footsteps on a floor would be heard mainly as impact sound in the room below and as airborne sound in the room above. See the *airborne and impact sound* diagram on page 19.

# Sound Insulation

Every designer and builder will find it useful to understand the main effects and principles of sound insulation described in Part One of this book. There is no need for anyone to be intimidated by decibels!

However, the technical measurement of sound insulation does becomes specialised, especially if you wish to satisfy Building Regulations or a Court of Law. Some of the following sections therefore can only summarise the main points and indicate where you can get more information.

## Sound transmission coefficient

The sound insulation of a partition depends on the amount of sound energy transmitted across the partition. *The transmission coefficient (T)* is a measure of the amount of sound energy transmitted through the partition. The amount of sound energy which passes through the partition is compared with the amount of sound energy arriving (incident) at the wall.

$$T = \frac{\text{Transmitted sound energy}}{\text{Incident sound energy}}$$

UNIT: none − a ratio.

For example, if 200 units of sound energy falls on a wall and 2 units passes into the room then the transmission coefficient is 20/200 or 0.01.

## Sound reduction index

The ear doesn't judge sound according to energy content, so the ratios of energy found in the transmission coefficient need to be converted to a decibel scale. The sound reduction index is used as a practical sound insulation value.

- *Sound reduction index* (SRI) is a measure of the insulation against the direct transmission of airborne sound.

The SRI is calculated by the following formula

$$SRI = 10 \log_{10} (1/T)$$

UNIT: decibels (dB).

Where:

$T$ = transmission coefficient for the partition under test.

The transmission coefficient varies with frequency and therefore the SRI also changes with frequency. The sound reduction index values for partitions, such as walls, need to be measured over agreed intervals of frequency; usually over 1/3 octaves between 100 Hz and 3150 Hz.

The multiple values of SRI can then be quoted for each frequency or, more usefully, presented on a graph with SRI values plotted against frequency values.

Where convenient, it is possible to compare SRI at a single frequency, such as those at 500 Hz. Typical values of sound insulation are given in Part One.

### Sound insulation calculation

A wall transmits 1 per cent of the sound energy which falls upon it at a certain frequency. Calculate the sound reduction index of the wall.

$$T = \frac{\text{Transmitted sound energy}}{\text{Incident sound energy}} = \frac{1}{100} = 0.01$$

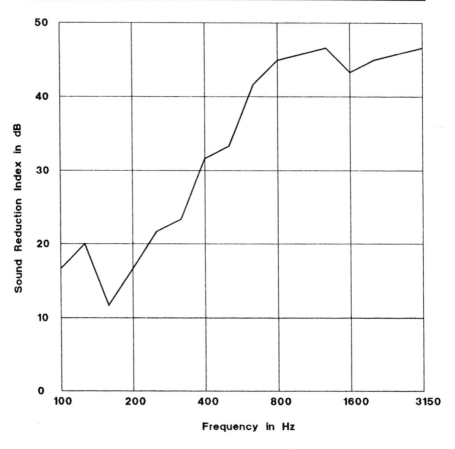

*Typical insulation curve.*

Using the formula

SRI = 10 log (1/$T$)

Substituting the known values into the formula gives

SRI = 10 log (1/$T$) = 10 log (100) = 10 × 2

SRI = 20 dB

## Composite partitions

Good sound insulation includes the principle of completeness or uniformity of a partition. Practical walls, however, often have to contain windows and doors which have a lower Sound Reduction Index than the wall. The SRI values of the different parts of the composite partition cannot be simply averaged according to areas.

But transmission coefficients can be averaged by area, using the following formula.

$$T_\mathrm{o} = \frac{(T_1 \times A_1) + (T_2 \times A_2) + (T_3 \times A_3)}{A_1 + A_2 + A_3}$$

Where:

$T_\mathrm{o}$ = overall transmission coefficient
$T_1$ = transmission coefficient of one component
$A_1$ = area of that component etc.

The overall transmission coefficient can then be used to calculate the overall sound reduction index.

## Measurement of sound insulation values

The SRI of a test partition can be measured in a specialist laboratory consisting of two rooms which are isolated so that no flanking paths are possible between them. All sound must pass through the test partition which is fitted between the two rooms.

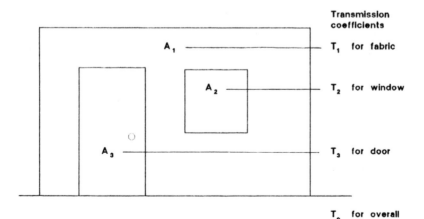

*Composite partition.*

*Field tests* can also be made on partitions which are installed in a building. These more realistic tests take account of actual conditions including the flanking transmissions in the building.

All measurements, whether in a laboratory or in the field, need to be made in a systematic manner. For the United Kingdom the test procedures are defined and detailed in British Standards, currently the document BS 2750.

The essence of all the tests is to generate an appropriate airborne or impact sound in the source room and to measure it in the source room and the receiving room over the agreed bands of frequency. Impact sound is generated by a 'standard tapping machine' which uses standard metal hammers or weights to make a 'standard impact'.

Some readings are repeated for different positions in the rooms and, in field tests, measurements must be made for least four pairs of rooms. The readings are then treated in a standard manner laid down by the British Standards.

*Airborne sound* test results are expressed as a *difference* in sound level between the source room and the receiving room. Higher figures mean better insulation against airborne sound.

*Impact sound* test results are expressed as the *actual* sound level in the receiving room when the sound source is a standard tapping machine in the room above. Lower figures mean better insulation against impact sound.

For both airborne and impact sound, the results have to be adjusted to take account of the acoustic effect of the receiving room. These adjusted values are known as the *standardised* values; in the past they have also been termed 'normalised' values.

For airborne sound insulation, standardised level differences (D) are quoted. For impact sound insulation, standardised sound pressure levels (L) are quoted.

## Sound insulation ratings

When sound insulation has been measured in accordance with the standard procedures described above, the result is a series of figures. One sound insulation result is generated for each of the agreed frequency bands and the set of results can be plotted as a curve like the one shown before. It is usually more convenient to express the set of results as a single figure but, because the decibel is a logarithmic number, it is not possible to use a simple numerical average.

An accepted method of generating a single number is laid down by a British Standard (BS 5281) and results in *weighted* index. The method uses a standard reference curve of sound insulation value plotted against frequency. The reference curve is placed over the test curve and moved up and down until the best fit is obtained in accordance with specified rules. The value of the reference curve at 500 Hz is then read and used as the single figure rating of sound insulation and called the *weighted* reading.

- Airborne sound insulation is quoted as a single Weighted Standardised Level Difference or $D_{nT,w}$
- Impact Sound insulation is quoted as a single Weighted Standardised Sound Pressure Level or $L_{nT,w}$

### Building Regulations

The Building Regulations for England and Wales are typical of building codes about sound insulation and they essentially limit the sound transmission between different dwellings. The statutory requirement is that specified walls and floors shall have 'reasonable resistance' to airborne or impact sound.

The particular walls and floors described below are specified by the current Building Regulations. As a future trend, it is expected that requirements for adequate sound insulation will be extended to include other types of buildings, for both new construction and for a change of use such as the conversion of an old house to flats.

### *Walls*

The following walls need to be sound-resistant against airborne sound only.

- Walls which separate a dwelling and another dwelling; or another building.
- Walls which separate a habitable room and any other part of the same building which is not in the same dwelling; or

machinery room or tank room; a place used for any other purpose unless it is only occasionally used for maintenance and repair.
- Walls which separate any part of a dwelling and a refuse chute.


### *Floors*

The following floors need to be sound-resistant against airborne sound.

- Floors which separate a dwelling *above* and another dwelling; or part of the same building which is not in the same dwelling including a machinery or tank room.
- Floors which separate a dwelling *below* and another dwelling; or part of the same building which is not in the same dwelling including a machinery or tank room.

The following floors need to be sound-resistant against impact sound.

- Floors which separate a dwelling *below* and another dwelling; or part of the same building which is not in the same dwelling including a machinery or tank room.

The Regulations can be satisfied by several methods. The simplest method is to use one of the widely used wall or floor constructions methods shown in the Approved Documents which accompany the Building Regulations. A selection of these constructions is also shown in Part Two of this book.

Another method of satisfying the Regulations is to repeat a type of wall or floor which has already been used and shown to be acceptable using the procedures for the measurement of sound insulation described earlier in this section.

# Building Sites

Part One of the book summarises some general principles for the control of noise, especially noise within buildings. This section focuses on noise in the construction industry and the problems it can give to the workers on a building site and to the environment surrounding the site.

## Planning and site preparation

Noise problems on building sites are best solved during the planning of a project and the setting up of the site and the trend to stronger health and pollution laws are a good incentive. A project should be planned so that the layout of the site, the machinery used and the number of building operations help to keep noise to a minimum.

Noise considerations should form part of formal tenders and method statements. Early consultation and negotiations with local authorities about limits and restrictions have both practical and legal advantages.

## Control of site noise

Where possible, the source of a noise is the best place to control the noise. Factors to be considered in the planning and operation of a construction include the following.

- *Site layout*. The position of access roads, materials stocks, batching plants, and generators all affect noise.
- *Machinery*. Noisy plant or processes should be exchanged for quieter alternatives.

- *Enclosures.* The effect of some noisy machines and processes can be made quieter by enclosing the machine or working within screens.
- *Duration.* Noisy operations are more annoying and damaging if they continue for longer periods.
- *Working hours.* Some periods of the day are more noise-critical than others, especially those that involve sleep disturbance.
- *Working methods.* Often, a quieter procedure is also good practice, such as lowering items rather than dropping them!
- *Distance.* Sound levels decrease as distance from a source increases. Sometimes there are options for maximising the distance between noise sources and noise-sensitive areas, such as housing.
- *Screening.* A barrier can give a reasonable reduction in noise levels. The positioning of site offices, materials stores and excavations can give useful noise reduction.

## Noise assessment

Noise levels generated on a typical construction site vary during the working hours of the day. The index used to measure such fluctuating noise levels is the Equivalent Continuous Sound Level, $L_{eq}$, described in an earlier section on Noise Measurement. $L_{eq}$ is a statistical record of the total noise energy or 'noise dose' received during a specified number of hours, and different 'patterns' of noise can give the same value of $L_{eq}$. This exposure to sound energy is found to correlate with hearing damage to people working with noise, using $L_{eq}$ (8 hours) = 90 dB(A) as a typical limit. The annoyance to environment surrounding construction sites can also be equated to the total sound energy received and $L_{eq}$ (12 hours) = 75 dB(A) is one limit that has been used.

Equivalent continuous sound levels are measured using a specialist sound level meter with recorder which automatically samples the noise at frequent intervals during the time period and calculates the $L_{eq}$. The noise exposure of individual employees can also be measured by noise 'dose meters' which are worn during work.

The *prediction* of likely noise levels on a construction site can be based on past records of site noise. In addition, acceptable estimates can be prepared from information published in British Standards and by bodies such as the Construction Industry Advisory Committee. These contain the $L_{eq}$ values for typical combinations of machinery, plant, operations, and on-times.

### Noise and the law

There is a general tendency to increase the effectiveness of laws relating to noise. For failure to comply with some laws, especially those concerned with safety, there are heavy fines, daily penalties and a personal criminal record.

The following review of noise and law is based on specific British legislation, but the general principles have wider applications. There is a helpful trend for many countries to base their technical standards, such as British Standards, on the widely agreed technical standards of the International Standards Organisations (ISO).

The British law relating to noise is contained within a large number of statutes (written laws), common law and legal precedent. Noise is treated as a type of legal *nuisance* known as a statutory nuisance which can be defined as 'an act or omission which has been designated a nuisance by statute'.

### Noise nuisance

The main statute or law regarding noise nuisance for England and Wales is the Control of Pollution Act 1974. As the title suggests, this Act of Parliament deals with a variety of matters affecting the environment.

The Act gives local authorities the duty and power to control noise from construction sites and similar sources of noise nuisance. The powers can be used either before or after the start of work. The local authorities can serve notices on contractors requiring that any nuisance be abated or prohibited.

A section of the Act entitles an individual to make a complaint to the Magistrate's Court which can result in an order to reduce the noise nuisance. The Act also allows contractors to initiate enquiries to the local authorities and to agree noise patterns before work starts. As well as being good practice, this consent restricts the actions which can be made against a contractor.

## Hearing damage

In addition to causing nuisance, noise can damage health. The Health and Safety at Work Act 1974 (The 'HSW Act') includes the protection of hearing from occupational noise and is relevant to work on construction sites which are now among the riskiest of occupations for hearing damage.

The Act generally requires all employers to ensure the health, safety and welfare at work of all their employees. The Act also requires designers, manufacturers, importers or suppliers to make sure that articles for use at work are designed and constructed to be safe and are provided with adequate information about their safe use.

In terms of occupational noise, the levels of noise exposure need to be kept below a recognised hazardous level which, at the moment, is an $L_{eq}$ (8 hr) of 90 dB(A). If the noise exposure is greater than this limit then other methods of hearing conservation, such as ear protectors, are needed.

# Product File

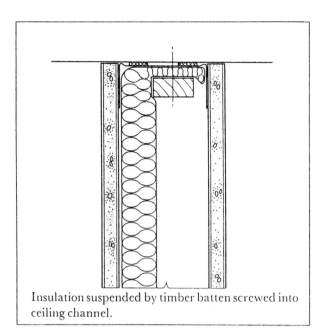

Insulation suspended by timber batten screwed into ceiling channel.

*Method for timber or plasterboard ceiling*

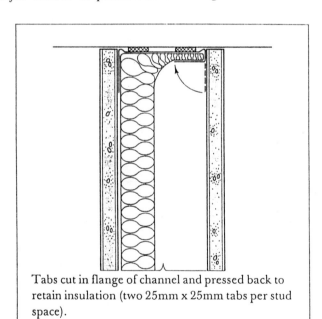

Tabs cut in flange of channel and pressed back to retain insulation (two 25mm x 25mm tabs per stud space).

*Method for concrete soffit, or similar*

*Sample applications of GYPGLAS 1200 glass wool used for sound insulation within partitions. Information available from Gyproc Insulation Limited, Whitehouse Industrial Estate, Runcorn, Cheshire WA7 3DP.*

INTERNAL PARTITIONS

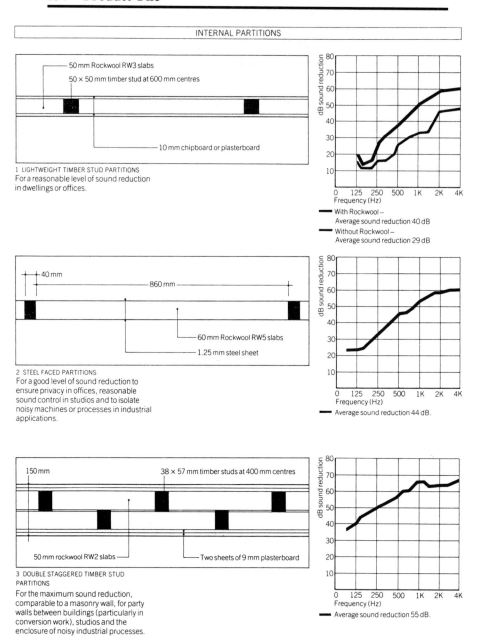

1 LIGHTWEIGHT TIMBER STUD PARTITIONS
For a reasonable level of sound reduction
in dwellings or offices.

50 mm Rockwool RW3 slabs
50 × 50 mm timber stud at 600 mm centres
10 mm chipboard or plasterboard

With Rockwool –
Average sound reduction 40 dB
Without Rockwool –
Average sound reduction 29 dB

2 STEEL FACED PARTITIONS
For a good level of sound reduction to
ensure privacy in offices, reasonable
sound control in studios and to isolate
noisy machines or processes in industrial
applications.

40 mm
860 mm
60 mm Rockwool RW5 slabs
1.25 mm steel sheet

Average sound reduction 44 dB.

3 DOUBLE STAGGERED TIMBER STUD
PARTITIONS
For the maximum sound reduction,
comparable to a masonry wall, for party
walls between buildings (particularly in
conversion work), studios and the
enclosure of noisy industrial processes.

150 mm
38 × 57 mm timber studs at 400 mm centres
50 mm rockwool RW2 slabs
Two sheets of 9 mm plasterboard

Average sound reduction 55 dB.

*Sample applications and performance figures of Rockwool
mineral wool products used for sound insulation within
partitions. Information available from Rockwool Limited,
Pencoed, Bridgend, Mid Glamorgan CF35 6NY.*

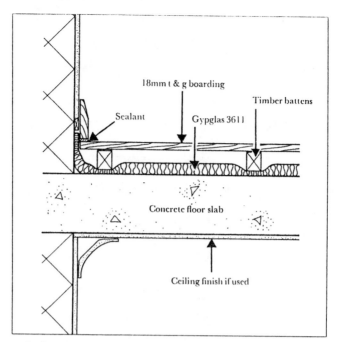

*Timber raft floating layer (showing floor/wall junction)*

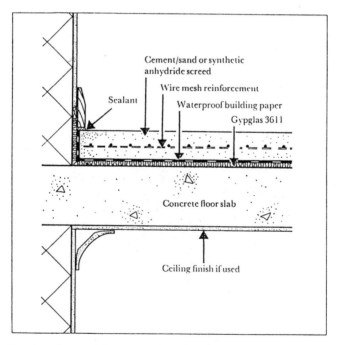

*Screed floating layer (showing floor/wall junction)*

*Sample applications of GYPGLAS 3611 glass wool used for sound insulation within floors. Information available from Gyproc Insulation Limited, Whitehouse Industrial Estate, Runcorn, Cheshire WA7 3DP.*

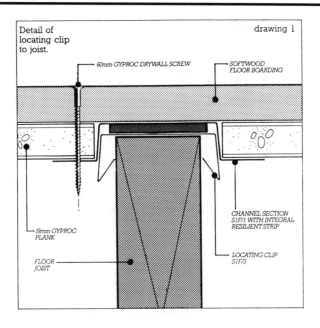

Detail of locating clip to joist.

drawing 1

60mm GYPROC DRYWALL SCREW

SOFTWOOD FLOOR BOARDING

CHANNEL SECTION S1F/1 WITH INTEGRAL RESILIENT STRIP

19mm GYPROC PLANK

FLOOR JOIST

LOCATING CLIP S1F/3

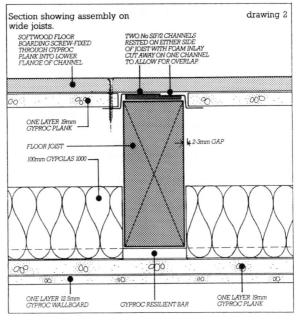

Section showing assembly on wide joists.

drawing 2

SOFTWOOD FLOOR BOARDING SCREW-FIXED THROUGH GYPROC PLANK INTO LOWER FLANGE OF CHANNEL

TWO No SIF/2 CHANNELS RESTED ON EITHER SIDE OF JOIST WITH FOAM INLAY CUT AWAY ON ONE CHANNEL TO ALLOW FOR OVERLAP.

ONE LAYER 19mm GYPROC PLANK

FLOOR JOIST

2-3mm GAP

100mm GYPGLAS 1000

ONE LAYER 12.5mm GYPROC WALLBOARD

GYPROC RESILIENT BAR

ONE LAYER 19mm GYPROC PLANK

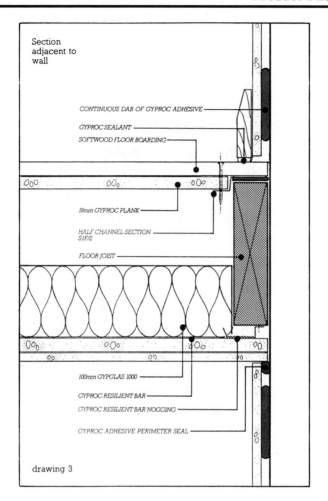

Section
adjacent to
wall

CONTINUOUS DAB OF GYPROC ADHESIVE

GYPROC SEALANT

SOFTWOOD FLOOR BOARDING

19mm GYPROC PLANK

HALF CHANNEL SECTION
S1F/2

FLOOR JOIST

100mm GYPGLAS 1000

GYPROC RESILIENT BAR

GYPROC RESILIENT BAR NOGGING

GYPROC ADHESIVE PERIMETER SEAL

drawing 3

*Sample details of Gyproc S.I. Floor System for remedial
treatment sound insulation of existing timber joist floors.
Information available from British Gypsum Limited, Church
Manorway, Erith DA8 1BQ.*

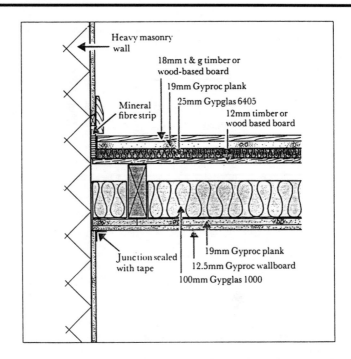

*Sample application of GYPGLAS 6405 and GYPGLAS 1000 used for sound insulation within timber PLATFORM floor. Information available from Gyproc Insulation Limited, Whitehouse Industrial Estate, Runcorn, Cheshire WA7 3DP.*

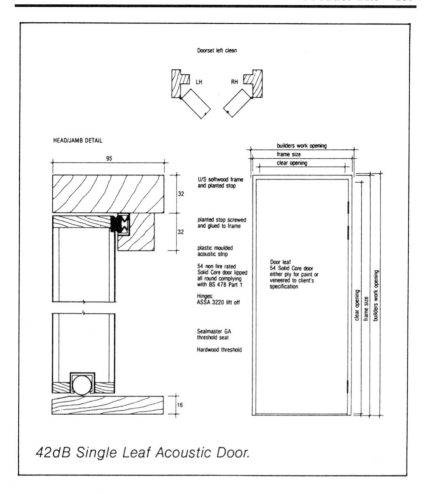

*42dB Single Leaf Acoustic Door.*

## Table 1 Acoustic door performance

| Model | Combined Wall Rating (dB) | Sound Insulation Index (ASII) (dB) | Average SRI (100–3150 Hz) (dB) | Average SRI (100–10 000 Hz) (dB) | Sound Trans. Class (STO) (dB) |
|---|---|---|---|---|---|
| 38dB | 38 | 31 | 28 | 30 | 31 |
| 42dB | 42 | 35 | 33 | 34 | 35 |

*Construction and performance details of a typical sound resisting door. Information available from Sound Attenuators Limited, Eastgates, Colchester, Essex CO1 2TW.*

REDUCING AIR, LIQUID OR PARTICLE-SOLID BORNE NOISE WITHIN DUCTS AND PIPES

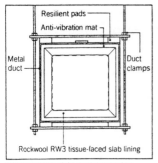

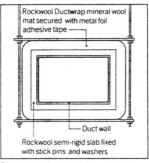

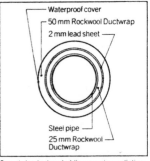

Internal acoustic lining to attenuate duct-borne noise

External acoustic casing to duct to reduce ambient noise levels

Sound-deadening cladding on noise-radiating large diameter pipe

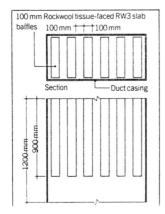

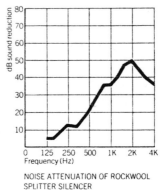

NOISE ATTENUATION OF ROCKWOOL SPLITTER SILENCER

*Sample applications and performance figures of Rockwool mineral wool products used for sound insulation within ducts and pipes. Information available from Rockwool Limited, Pencoed, Bridgend, Mid Glamorgan CF35 6NY.*

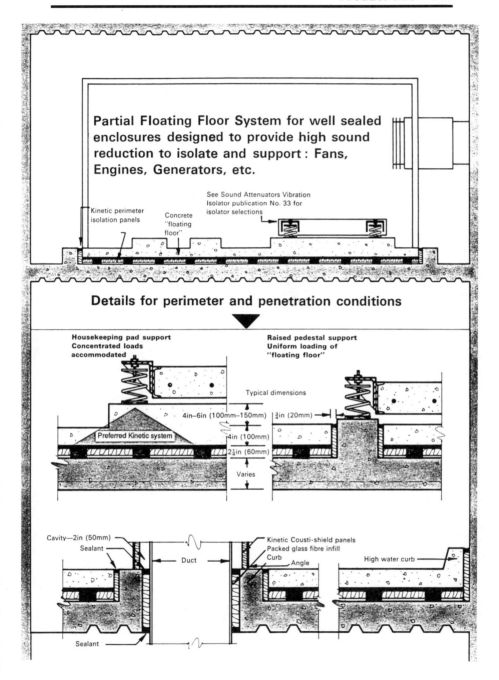

Partial Floating Floor System for well sealed enclosures designed to provide high sound reduction to isolate and support: Fans, Engines, Generators, etc.

See Sound Attenuators Vibration Isolator publication No. 33 for isolator selections

Kinetic perimeter isolation panels

Concrete "floating floor"

**Details for perimeter and penetration conditions**

Housekeeping pad support
Concentrated loads accommodated

Raised pedestal support
Uniform loading of "floating floor"

Typical dimensions

Preferred Kinetic system

4in–6in (100mm–150mm)
4in (100mm)
2¼in (60mm)
Varies

¾in (20mm)

Cavity—2in (50mm)
Sealant

Duct

Kinetic Cousti-shield panels
Packed glass fibre infill
Curb
Angle

High water curb

Sealant

*Sample details of a typical noise isolation system for mechanical equipment. Information available from Sound Attenuators Limited, Eastgates, Colchester, Essex CO1 2TW.*

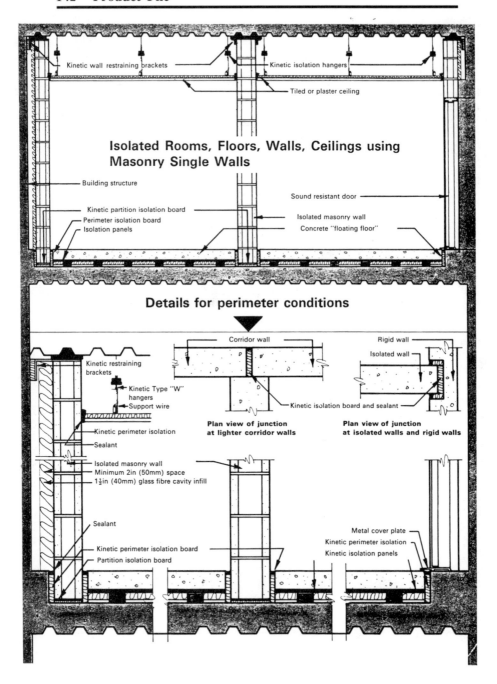

*Sample details of a typical insulation system for an isolated room.
Information available from Sound Attenuators Limited,
Eastgates, Colchester, Essex CO1 2TW.*

# Index